# For Love of the Earth

Francis H. Chapelle

# DEDICATION

For Topher, a son to be proud of.

# CONTENTS

# ACKNOWLEDGMENTS

We gratefully acknowledge our professors at the University of Maryland.
Galt Siegrist, Ann Wylie, Peter Stifel, and Jerry Weidner. Thank you for
instilling in us a love for the Earth.

# CHAPTER 1.
## A MEANINGFUL ROCK

In the summer of 1982, a young man of about 30 years of age happened to take a walk along the shore of Lake Huron in Michigan. Lake Huron had been carved out of the local limestone bedrock by glaciers during the ice ages. As such, the shores are littered with boulders and stones that were carried by the grinding ice and left behind as the glaciers melted. As he walked along, a particularly colorful rock drew his attention. It was about the size of a bowling ball and was a dark gray, fine grained rock that had been rounded and smoothed by the lake's wave action. But what made it colorful were the gleaming chunks of bright red granite embedded in the black matrix (Figure 1).

Figure—1. The mysterious rock from Lake Huron. Pocketknife is for scale.

Curious, he picked the rock up and examined it more closely. The black matrix was a garden-variety shale, a sedimentary rock that is often formed in deep ocean water. But the chunks of red granite, which varied in size from a couple of millimeters to a few centimeters, looked as freshly broken as if they had just come from a quarry. How is it possible for a sedimentary rock deposited in deep water to have chunks of angular granite, a crystalline igneous rock, embedded in it? The young man, a geologist, didn't have a clue. Puzzled, he picked up the mysterious rock and carried it home. When he got there, he showed it to his wife who was also a geologist. She looked at it closely, turning it over and over in her hands. Finally, she just shrugged. She had no idea of what it could be. But, making the best of it, she remarked that it would look nice in their garden. So they put it in their garden where it resides to this day.

For years after that, the young man would occasionally see that rock in his garden and wonder about it. Then, in 1998 he happened on a paper that had been published in the journal *Science* entitled "A Neoproterozoic Snowball Earth".[1] Intrigued, he read the paper which argued that the climate had abruptly cooled 700 million years ago, leading to a world-wide glacial event that had covered most of the Earth with ice. Then he remembered the mysterious rock he had found in Lake Huron. Could that be the answer to the puzzle of the rock?

It turns out that several times in the history of the Earth, there have been episodes of extreme cold that led to extensive world-wide glaciation[2]. The oldest of these events, known as the Huronian glaciation, occurred between 2.4 and 2.1 billion years ago and may have been triggered by the appearance of photosynthesis on Earth. One side effect of photosynthesis is to remove carbon dioxide from the atmosphere. The depletion of carbon dioxide—a heat-trapping greenhouse gas—from the atmosphere may have initiated catastrophic cooling, resulting in a "Snowball Earth".

The Huronian glaciation, so named because the rocks recording that episode are exposed in Canada just north of Lake Huron, covered much of the Earth with glaciers. As glaciers advance, they plow and break up the rocks in their way. These angular, broken chunks of rock get picked up by the ice and are carried along by the advancing glacier. When the glaciers finally reach the open sea, they splinter off forming huge icebergs that then float out over the ocean. As the icebergs slowly melt, the angular pieces of rock—in this case red granite from the Canadian interior—are released and drop to the seafloor. If these rocks (called "dropstones") happen to fall onto the fine-grained clays and silts typically found on the seafloor, and if those sediments are subsequently turned into a black shale, you get the rock the young man found on the shores of Lake Huron. In fact, the rock he had found came from the 2.2 billion year-old Gowganda Formation that crops out at land surface in Canada.[3] It had been gouged out by glaciers during the most recent ice age, carried south to what eventually became Lake Huron where it was unceremoniously dumped. So, sixteen years after he picked up what at first was just a colorful, interesting, mysterious-looking rock, the young man finally knew where it came from and how it had come to be.

Most people, like the young man on the shores of Lake Huron, collect rocks because they are unusual in some way. They might contain pretty crystals, or swirling bands of color, or fragments of fossil snails and clams. And, like the young man's mysterious rock, they often end up in someone's garden or on a living room mantelpiece. But while rocks can be attractive objects, there is always an underlying—and often hidden—reason for their unique appearance. That is because every single rock in the world *records something that happened in the history of the Earth.* And if that something happens to be significant, such as a two-billion year-old episode when most of the Earth was covered in ice, then that rock becomes more than just a pretty object.

It becomes meaningful.

----------------

For most of human history, rocks and minerals were viewed primarily as objects that were more or less useful. Two million years ago, our earliest ancestors of the genus *Homo* discovered that fist-sized rocks made of chert could be chipped into crude hand axes and used to crack nuts or split bones. Over the millennia people learned that stones could be used to build shelters, that gold pebbles found in streambeds could be fashioned into jewelry, or that chunks of native copper could be pounded into tools. Viewing rocks as objects, some of which are useful and most of which are not, is still how most people think about rocks today.

But beginning in 17<sup>th</sup> century Italy, something different began to happen. A particularly observant physician, walking in the countryside of Tuscany, happened to notice that shark's teeth were embedded in the local limestone [4]. He reasoned that that could only mean the limestones had been deposited in an ancient sea. Furthermore, since sediments in the sea are deposited continuously, he also reasoned that younger sediments must overlie older sediments. That led to the astonishing idea that rocks could *record the passage of time.* Could it be that ordinary rocks, which most people scarcely notice in their everyday lives, had a meaning deeper than just what they could be used for? Could it be that rocks are a window into the dim, misty, hidden past of the Earth?

The young man who picked up that mysterious rock on the shore of Lake Huron, and his wife to whom he showed it, were clearly of the mindset that the rock had to mean *something* if they could only discover what it was. To them the fact that the rock was colorful and pretty was secondary to the notion that it represented something unusual that happened long ago. So the question is, how did they acquire that mindset? How is it that some people, like the young man and his wife, who from childhood naturally began with the static mindset of rocks as objects, become aware that rocks actually record a historical event?

This book is partly the story of how humans slowly, tentatively, and sometimes painfully learned to read the story of Earth's past that is written into rocks. That it took well over two hundred years for the idea of rocks as history to be established is unsurprising. It took time for people to get past the static view that rocks are significant only for what they can be used for. But it's also the story of how two people who, also over a period of many years, experienced that same mindset change. Beginning when they were both very young, these two people began to take notice of the rocks around them interesting objects. Noticing led to curiosity, and that curiosity drew them to study geology when they got to college. As they worked their way through the curriculum, and later when they got their first jobs, their world-view changed, bit-by-bit, from rocks as objects to rocks as meaning.

It was an awakening to the Earth.

## REFERENCES

1. Hoffman, P.F., Kaufman, A.J., Halverson, G.P. and Schrag, D.P., 1998. A Neoproterozoic snowball earth. Science, 281(5381), pp.1342-1346.
2. Harland, W.B., 1964. Critical evidence for a great infra-Cambrian glaciation. Geologische Rundschau, 54(1), pp.45-61.
3. Miall, A.D., 1983. Glaciomarine sedimentation in the Gowganda Formation (Huronian), northern Ontario. Journal of Sedimentary Research, 53(2): 477-491.
4. Steno, Nicolas, 1669. Preliminary discourse to a dissertation on a solid body naturally contained within a solid.

# CHAPTER 2.
## CURIOSITY AND SELF INTEREST

The young boy, who had barely turned six, scrambled up the steep slope and paused to catch his breath. The woods around him sighed as the wind blew through the trees, giving him some welcome relief from the shimmering summer heat. The slope faced to the east, and the granite boulders that littered the slope around him had been thoroughly baked by the morning sun. Heat radiated from the rocks, scattering the sunlight and giving a faint glow to the surrounding landscape.

This particular patch of woods lay behind a row of stately brick houses that served as quarters for the senior officers stationed at the United States Military Academy at West Point, New York. It remained a patch of woods largely because the slope of the hill was so steep that nothing could be built on it. In fact, when the officer's housing had been built sometime in the early twentieth century, the granite had to be drilled and blasted just to make room for the houses. But in any case, the rocks that were scattered on the hillslope seemed so ordinary that the boy had never paid any attention to them.

But this summer morning, the boy was on a different mission. He had just found out, probably from watching TV, that the Mohican Indians who had once lived in these woods had used stone tips for their arrows. That sounded cool. If Indians could make stone-tipped arrows, then he could too. The hill behind his father's quarters (15B) included several steep cliffs of granite (Fig. 2.1), and the ground was littered with rocks and loose stones. It was the obvious place to look for stones that he hoped to chip into arrowheads.

Figure 2.1—A cliff face of the granite behind the young boy's house in West Point, NY. Note silvery color imparted by crystals of muscovite (mica) in the rock below the pocketknife.

So, for the first time the boy actually paid attention to the rocks that littered the slope behind his house[1,2]. At first he was disappointed. The granite rocks were certainly hard, and if you banged them together you could break them into sharp-edged pieces. But the shape of the resulting pieces were so irregular that it soon became evident that shaping them into arrowheads was going to be a difficult chore. The boy had never been fond of difficult chores. But while he was banging the rocks together, he noticed that they contained bits of shiny grains that looked like silver. He knew from watching TV that the Lone Ranger used bullets made out of silver, and that silver therefore must be very valuable. I'm rich, he thought, I've discovered silver. Hastily he gathered up several specimens of the richest-look silver "ore" and headed back down the hill.

For several days after this the boy hoarded his new find in secret. His dilemma was that if anyone found out about his discovery of silver, they might somehow steal it away from him. On the other hand, he had absolutely no idea how to go about extracting the silver from the rock so he could sell it. Finally, he decided to ask the person who generally knew the answer to everything—his Mom. Waiting until she was alone—no need to let anyone else in on the secret—he showed her the rocks with the silvery-looking mineral and asked if it really was silver. Somberly his mother examined the rocks, turning them over and holding them up to the light.

"Sweetheart", she began, "I don't think this is really silver. I think it's just ordinary mica." Even though the boy didn't know what mica was (Fig. 2.1), he knew it wasn't the silver riches that he was hoping for.

And he was crushed.

----------------

Trying to make arrowheads out of stone, or looking for silver ore in chunks of granite may seem childish. But it's not. In fact, you can seriously argue that using rocks for tools, or rocks as building materials, or rocks as sources of gems or metals, is a defining characteristic of the human species. Thanks to modern anthropology, that rock-utilizing behavior can be traced back to the very earliest humans.

In 1935, Louis Leakey and his new wife Mary returned to Kenya, the country of his birth, with the goal of excavating fossils of our human ancestors. Louis had formal training in anthropology but Mary did not. Mary did have, however, what would prove to be a much more useful skill for what the couple eventually would accomplish—she had been trained in the techniques of scientific archeological excavation.

Just how she acquired those skills is a story in itself. Mary had been sent to a Catholic convent school in England at age 13 after her father had died of cancer. In a word, Mary hated school. She was expelled from her first school for refusing to recite poetry. She was expelled from her second school for causing an explosion in the schools' chemistry lab. With a school record like that, Mary was never going to be admitted to any college in England. But Mary was also a talented artist and illustrator. At age 17, after surreptitiously attending a few lectures in archaeology at a local college, she began working as an illustrator for summer archaeological excavations being carried out in England. In addition to learning the art of archaeological illustration, over the next four years she also absorbed the techniques of quantitative archaeological excavation. She learned how to set up grids at each site so the locations and context of individual artifacts could be accurately cataloged, and she learned how to excavate in layers so that the site's stratigraphy could be studied, recorded, and analyzed.

Louis Leakey, for all his academic training in anthropology, knew none of this. But because Mary insisted, and later as their value dawned on Louis, they used these techniques while excavating various sites. When the Leakeys began excavating at Olduvai Gorge in the 1940s, Mary immediately noticed the presence of stone tools that appeared to be some sort of hand axes. The stratigraphy of the site, which Mary had carefully documented, suggested they might be as old as two million years.

But who had made them? It took more than ten years of work at Oldavai for the Leakeys to find a few teeth, a lower jaw, and a partial cranium of a creature that had distinctly human traits. The teeth were important because they clearly belonged to an individual that was less robust than other hominin fossils they had found previously at Oldovai. The cranium was important because it suggested that the brain case was relatively large.

After the discovery of similar fossils in other locations, Louis Leakey published their findings in 1964, proposing that these fossils were the first ever to be found of the genus *Homo*. [3] Furthermore, because of the stone tools found with the fossil bones, they named it *Homo habilis* which is Latin for "handy man". That is, a man who could make stone tools. Louis never admitted that he took the stone tools into account when deciding that the fossils represented the earliest known humans, but he clearly did.

So, making tools out of stone like the young boy from West Point was trying to do, was not just idle childish play. It is one of the fundamental behaviors that made us human in the first place.

----------------

For most of human history, people like our *Homo habilis* ancestors with their hand axes, were primarily interested in rocks and minerals for whatever practical use they could be put to. Even after Leakey's 1964 publication, it was a bit of a mystery as to what the hand axes were actually used for. They couldn't be for chopping firewood because *H. habilis* had not yet learned to use fire. Years later, some studies suggested that *H. habilis* was using the axes as percussion tools[4], perhaps to break open nuts or to splinter bones in order to reach calorie-rich bone marrow. Alternatively, sharp edged flake chipped from the axes could have been used to slice meat from scavenged bones[5]. But regardless of what they were used for, those tools apparently gave our ancestors a certain advantage when competing with other species of plant gatherers and scavengers living near the Serengeti Plain of Africa.

So from the very beginning of humanity, people used rocks for tools, stones for building, ores for smelting copper, tin, and iron, coal for fuel, and gems for jewelry. In each case this was driven exclusively by self-interest. However, while self-interest determined how rocks and minerals were actually used, their discovery was often the result of simple human curiosity. But even when people began noticing rocks out of curiosity, things often circled back to self-interest.

Take, for example, the story of James Hutton.

---------------

Modern scientific geology often traces its beginnings to the life and studies of James Hutton (1726-1797). Hutton was born in Edinburgh, Scotland, and was the son of a relatively prosperous merchant and property owner. His mother saw to it that James attended the High School of Edinburgh, at the time the best secondary school in Scotland. He went on to the University of Edinburgh where he began to attend lectures in medicine. He finally graduated as a Doctor of Medicine from the University of Paris in France.

But rather than practicing medicine, Hutton returned to Scotland to manage the two farms that his family owned. Wanting to make the farms more profitable, Hutton set about clearing trees and draining swampland in order to increase his productive acreage. In the process of doing this, he began to take notice of the rock formations that underlay his land. In 1753, while still in his 20s, he wrote to a friend that he had:

> ....become very fond of studying the surface of the
> earth, and was looking with anxious curiosity into
> every pit or ditch or bed of a river that fell in his
> way.

In other words, self-interest (having a more productive farm) led to him to be genuinely curious about the earth, and more specifically how the earth had been made.

Hutton started taking long walks in the Scottish countryside to look at rock formations, wondering how they might have come to be. Over the next several years, he noted in particular how rock formations related to each other spatially and how they exhibited a consistent stratigraphic succession. This work, as the above quote clearly shows, was entirely curiosity-driven. But it wasn't long before Hutton's self-interest reappeared.

As far back as the 17th century, there had been talk of building a trans-Scotland canal. The seas around Scotland are notorious for their storms and reefs, and moving goods by way of ships and boats was a risky business. Having an inland canal would be much safer, and thus more profitable. But digging canals (by hand) was an expensive undertaking, and so it wasn't until 1768 that Parliament authorized the construction of a canal between the Firth of Forth clear across Scotland to the Firth of Clyde. What Parliament didn't do was appropriate funds for the canal's construction. That little detail was left up to private investors. It took a while to assemble enough investors, but they eventually did and formed a company called "The Company of Proprietors of the Forth and Clyde Navigation". In what would now be called an "initial public offering", 1,500 shares were offered at £100 per share. Some of those shares, it's not clear just how many, were purchased by one James Hutton.

For the next seven years, Hutton was closely involved in the construction of the Forth and Clyde Canal. His knowledge about the rocks underlying the Scottish central lowlands, where the canal was built, was unprecedented. That knowledge was a tremendous help to the engineer, one John Smeaton (who first coined the term "Civil Engineering") during the canal's design and construction. But digging the canal was also a tremendous help to Hutton's geological inquires, since it effectively exposed rocks previously hidden from view.

Over the years, Hutton gradually came to the realization that the sedimentary rocks he observed could not have, as the bible said, been formed in a few days or in a single flood. He also realized that the Paleozoic rocks his canal exposed must have been deposited in the sea. In his own (somewhat opaque) words:

*The solid parts of the present land appear in general, to have been composed of the productions of the sea* (that is, sediments) *and of other materials similar to those now found upon the shores.*

And, since the rocks formed from these sediments were now dry land, it followed that:

*The consolidation of masses* (sediments) *formed by those tides and currents…and the elevation of those consolidation masses from the bottom of the sea, the place where they were collected, to the stations in which they now remain above the level of the ocean.*

Those ideas were revolutionary in the 18th century. For virtually all of human history, the significance of rocks and stones had been entirely economic—what practical use could they be put to? But what Hutton was saying, in his admittedly clumsy way, was that the real significance of rocks is what they revealed about the history of the Earth. That was an astonishing insight and was, with the exception of some earlier Italian geologists, a new thing under the sun.

Hutton went on to write a book, published in 1785, entitled "Theory of the Earth: or an Investigation of the Laws observable in the Composition, Dissolution, and Restoration of Land upon the Globe". In it he laid out what would become the founding principles of scientific geology. Among these was the realization of just how old the Earth must be. With considerable and uncharacteristic eloquence, he concluded a paper published in 1788 with the words:

*The result, therefore, of our present enquiry is, that we find no vestige of a beginning, and no prospect of an end.*

So, beginning at least with James Hutton, scientific geology acquired a duality that governs the discipline to this day. On one hand, it functions to observe the composition, distribution, and association of the rock formations that make up the earth, and to read the history of how, when, and why they were formed. On the other hand, as James Hutton did when he invested in the Forth and Clyde canal, it is to use that understanding to practical and profitable ends. From its very beginnings, it seems, scientific geology has been a marriage of self-interest and curiosity.

----------------

While the young boy from West Point had been bitterly disappointed that the silvery mica in the local granite was not in fact silver, he might have taken solace from the fact that that mistake had been made before. In 1862 an aspiring silver miner in Nevada—who would later take the pen name Mark Twain—did exactly the same thing. In his book *Roughing It*[6], he recounted this story:

> *I confess, without shame, that I expected to find masses of silver lying all about the ground....I crawled about the ground, seizing and examining bits of stone, blowing the dust from them or rubbing them on my clothes, and then peering at them with anxious hope. Presently I found a bright fragment and my heart bounded! The more I examined the fragment the more I was convinced that I had found the door to fortune.*

Taking his discovery back to camp, Mark Twain showed it to his fellow would-be miners.

> *"Gentlemen," said I, "...all I ask of you is to cast your eye on that, for instance, and tell me what you think of it!" and I tossed my treasure before them.*
>
> *There was an eager scramble for it, and a closing of heads together over it under the candle-light. Then old Ballou said:*
>
> *"Think of it? I think it is nothing but a lot of granite rubbish and nasty glittering mica that isn't*

*worth ten cents an acre!"*
  *So vanished my dream. So melted my*
*wealth away. So toppled my airy castle to the earth*
*and left me stricken and forlorn.*

  The young boy from West Point had been similarly stricken and forlorn. But he would later remember that it was the first time he had ever taken any notice of rocks.

## REFERENCES

1. LaMoe, J.P. and Mills, R.W., 1988. Field guide to the geology of West Point.

2. Johnson, M.C. and Gellasch, C.A., 2004. Geology in the West Point, New York region, and its influence on three centuries of local land use. Northeastern Geology and Environmental Sciences, 26, pp.285-297.

3. Leakey, L.S., Tobias, P.V. and Napier, J.R., 1964. A new species of the genus Homo from Olduvai Gorge. Nature 202:7-9.

4. Mora, R. and De la Torre, I., 2005. Percussion tools in Olduvai Beds I and II (Tanzania): implications for early human activities. Journal of Anthropological Archaeology, 24(2), pp.179-192.

5. Diez-Martín, F., Sánchez, P., Domínguez-Rodrigo, M., Mabulla, A. and Barba, R., 2009. Were Olduvai hominins making butchering tools or battering tools? Analysis of a recently excavated lithic assemblage from BK (Bed II, Olduvai Gorge, Tanzania). Journal of Anthropological Archaeology, 28(3), pp.274-289.

6. Twain, Mark, 1872. Roughing It. American Publishing Company, Hartford, Connecticut.

# CHAPTER 3.
## THE ROCK OF MATAPAN

It was 2:30 AM when the father shook the young girl awake. "Time to go, sweetheart."

The seven-year old little girl came awake slowly and sat up in bed. It was sometime in the 1960s, and every summer her family would drive from their home just outside Washington D.C. to Boston and visit their grandparents. Her father always liked to leave early, partly because of the notoriously bad traffic around Washington and New York, but also so they could catch the Staten Island Ferry when they got to New Jersey. Taking the ferry didn't save much in the way of driving time, but it did break up the dreary 9-hour trip and allow the children—the young girl, her older sister, and two younger brothers—to get out of the car and stretch their legs. It also gave them a chance to see the Statue of Liberty.

Actually, the young girls' grandparents didn't live in Boston itself. Rather, they lived in the town of Mattapan a few miles south of Boston. Upon arriving in Mattapan at two or three in the afternoon, the kids would say the obligatory hellos to their grandparents, and, because they were restless after the long boring trip, they were shooed out of the house to play.

"Go climb The Rock", their father would say.

The Rock was a curious feature. It was (is) located in downtown Mattapan just off River Street near the Neponset River. The Rock is greenish-gray in color, shaped like a dome, and it is huge—about the size of a football field—and is more than two stories tall. A picture of part of The Rock, as seen from the grandparent's house, is shown in Figure 3.1. When the house was built in the 1920s, the builder's plan was to build seven or eight houses on the west side of Suncrest Road. The Rock loomed on the east side of the road. Once the first seven or eight houses were built and sold, and the builder had replenished his working capital, he planned on dynamiting The Rock, clearing away the shattered

Figure 3.1—A (small) portion of The Rock of Mattapan, as seen from the grandparent's house.

debris, and building more houses where The Rock had been.

But the plan was a failure. When the builder drilled holes in The Rock and placed his dynamite charges, it proved to be so hard and tough that the dynamite couldn't break it up. The builder had better success when he tried using more powerful charges of dynamite. Unfortunately, he also managed to break most of the windows of the houses on the west side of the street, which he subsequently had to replace at his own expense. The only way the builder was going to destroy The Rock was if he also destroyed the houses he had already built. So he simply gave up. No more houses were built, and The Rock remains there to this day.

Doubtless this was bad news to the builder, but it was a boon for the local kids. The Rock was a made-to-order playground. The grandparents with their 3-year old son, later to become the young girl's father, moved into their house in 1929. And as he grew, the Father became accustomed to scrambling around and climbing on The Rock. So, it was natural that when his own children came to visit their grandparents for the summer, they too would spend a good deal of time climbing around on The Rock.

If The Rock had simply been a smooth dome, it would have been impossible to climb. The sides, while not vertical, were still pretty steep. But, even though The Rock was extremely hard, as the builder had discovered, it also contained a network of fractures that cut across at various angles to the ground (Fig. 3.1). The fractures were convenient hand- and foot-holds for the kids. There was one fracture in particular, located right across the street from the grandparent's house that you could reach with a single step. This particular fracture rose up at gentle angle, and so by shuffling sideways with your feet, and by holding on to other fractures with your hands, you could climb sideways right up to the top. The top of The Rock itself was fairly flat, and so once up, you could easily and safely (sort of) walk around. And it was cool to be able to look down at the roof of their grandparent's house.

While the little girl and her siblings played hide-and-seek on and around The Rock, she couldn't help but notice that there were several odd things about it. First of all, while its color was generally greyish-green, parts of it were actually pinkish in color. Secondly, if you looked closely at a freshly-broken piece of The Rock, you could see it was made of tiny crystals, some of which were red (giving the rock its pinkish color) but others were rectangular in shape and white in color. As the little girl daydreamed through the lazy summer days, she remembers wondering what those little white crystals might be.

----------------

Rocks have been important in the history of the Boston area from its very beginnings. The Puritans, who with the sponsorship of the Massachusetts Bay Company, set out for the New World in the spring of 1630. They originally were headed for the tiny settlement of Salem, fifteen miles north of Boston. Of the 1,000 or so Puritans that made this trip, most of them had been farmers in southern England which is known for its rich, fertile farmland. Upon reaching Salem, however, they were dismayed to find that the ground, like Mattapan, was covered with tough, hard rocks that yielded only poor, thin soils[1]. The prospects for farming weren't good, and the Puritans weren't happy about it. They needed a better spot to settle.

So, a week after their arrival in Salem, John Winthrop, the Governor of the Massachusetts Bay Colony, trekked southward over Indian trails to try and find a more auspicious place for the Puritans to settle. After walking for a few miles, Winthrop noticed that the landscape was sloping downward and that the rocky terrain gradually vanished, to be replaced by mossy forests and hills covered by thick soils. He didn't know it, but he had just descended into what would later come to be called the Boston Basin which, by virtue of its lower elevation, was largely covered by glacial drift left over from the ice ages. The hills near Boston, including Bunker Hill, Breed's Hill, and Beacon Hill, are glacial drumlins that, for reasons nobody quite understands, are unusually symmetrical. Even better, there was a river (the Charles) that fed into a wonderful natural harbor.

Upon Winthrop's return to Salem, the Puritans packed up their ships and sailed south to Boston Harbor. At first, they tried to settle on what is now the Cambridge side of the Charles River. But it soon became evident that the water supply there was tainted and unhealthy. Across the River an Englishman named William Blackstone, who had started a farm on one of the hills, informed the Puritans that there were several good-quality springs in the vicinity of Beacon Hill[1]. So, the Puritans moved across the Charles River and began to build a town that they named Boston, after an (obscure) town in England.

The rocky terrain that surrounds the Boston Basin, which the Puritans summarily rejected as a fit place to settle and farm, and the more pleasant terrain within the Boston Basin itself, are the result of a sequence of events that have occurred over a long, long period of time.

And it all started with The Rock of Mattapan.

---------------

Six hundred million years ago, as is still the case today, more than two thirds of the Earth's surface was covered by oceans. One difference between then and now, however, was that much of the land that did exist was clumped together into a supercontinent that we now call Gondwanaland or just Gondwana[2]. Gondwana later separated into what are now several different continents including Africa, South America, Antarctica, India and Australia. The process of that separation, however, involved much grinding and shifting of tectonic plates at the margins of Gondwana. That, in turn, resulted in seafloors being thrust deep beneath the landmasses, subsequent melting of rocks and sediments, all of which caused volcanoes to erupt. These weren't the relatively quiet, passive volcanos like we now see in the Hawaiian Islands, but the more explosive variety characteristic of the Northwestern United States or Indonesia. The reason they are so explosive is that as sediments are eroded and accumulate on the margins of continents, they become progressively enriched in silica ($SiO_2$). Quartz, the most common form of silica, is tough, highly resistant to erosion and thus tends to be the principal material left over in the highly weathered sediments that form beaches.

Anyhow, when these weathered sediments are buried and subducted beneath a continent and melted, they form a silica-rich magma which is very viscous. So viscous that it tends to plug the throats of volcanos, letting a lot of pressure to build up behind. At some point the pressure builds up to a critical point and the volcano explodes. That's what caused the explosion of Mount St Helens in Washington State in 1980 (Figure 3.2). These explosions can be huge, as much as a thousand times larger than that of Mount St Helens, forming enormous clouds of immensely hot ash and pumice. When the ash falls back to Earth, the heat welds the ash grains together forming hard, dense rocks called welded tuffs. But, since the sides of the volcanoes are often very steep, these welded tuffs erode, wash down the sides of the volcano, and are dumped into the ocean. If you look closely at The Rock, you can see it is composed of a jumbled mass of rock chunks, some of them rounded by transport in water and some of them not rounded, and all welded together by a pink glassy matrix (Fig. 3.3).

That, very briefly, is what formed The Rock of Mattapan six hundred million years ago. So, you have grinding tectonic plates and massive explosive volcanos, all of which makes for a dramatic story. But the even more interesting story is about *how* humans managed to figure all of that out.

Figure 3.2.—Mount St. Helens erupting in 1980. U.S. Geological Survey file photo.

Figure 3.3—Close-up of the Rock of Mattapan, with a pocket knife for scale. Note the large water-rounded clasts of volcanic rock and also chunks of angular non-rounded volcanic rock embedded in a fine-grained pinkish matrix. All of which are characteristic of volcanic breccias.

You could argue that untangling that history began when the Puritans noticed that the hard granitic rocks surrounding the Boston Basin were not particularly promising for productive agriculture. But exactly what those rocks were was not at all obvious. Many of the early descriptions of the rocks around the Boston area focused simply on describing what they looked like. A treatise published in 1880 by a geologist named William Crosby goes on for page after page simply describing what the rocks look like and what minerals they contain. [3] In describing the rocks he saw near Mattapan, he writes:

> *The belt of country one to two miles wide extending E. by N. Hyde Park to the outer end of Squantum, including the valley of the Neponset, ... is very complicated stratigraphically...and the petrosilex and breccia north of the Neponset.....*

He goes on to describe the "petrosilex" rock:

> *"..... the petrosilex....always presents a compact, grayish or greenish-white base, porphyritic with feldspar crystals, and the most of the rock is elvanite, holding grains of transparent quartz in addition to the crystalline feldspar.*

Note that while he goes into great detail as to what the petrosilex rocks look like, he doesn't have much to say about how they might have been formed. That's probably because he wasn't sure.

But that changed over the next 25 years. By 1905, Crosby had deduced that the rocks were in fact volcanic in origin, and he wrote:[4]

> *"It is not improbable that some of the andesite dikes have formed effective* (volcanic) *vents. But of unequivocal or normal necks there are not indications in the sedentary zones of the batholite or in the vicinity of the felsite necks; but they are to be found farther east, in the effusive felsites, the clearest examples occurring on either side of the*

*Neponset (River), in the Mattapan district of*
*Dorchester and the Columbine district of Milton."*
In other words, Crosby had progressed from simply describing
what the rocks *looked like*, to interpreting how they had actually
been *made*. They were, in fact, formed by active volcanos that left
"felsite necks" (the throats of ancient volcanos) and "effusive
felsites" (volcanic ash and breccia deposits).

The name "Mattapan Volcanic Complex" was coined by
two geologists named Laurence La Forge and B.K. Emerson in
1917. La Forge described the rocks as[5]:

> *an extensive series of flows, volcanic breccias, and*
> *accompanying pyroclastic sedimentary beds with*
> *and to some extent cut by, intrusive felsites and*
> *granophyric rocks.*

So, by 1917, the actual origins of these rocks were pretty well
established. What wasn't known was how *old* the rocks were and
*where* they originally formed. Emerson speculated, based on
stratigraphic position, that the rocks might be about 350 million
years old. But he didn't really know, and he said so.

The first radiometric method for estimating the absolute
age of rocks was based on the radioactive decay of uranium to
lead. As early as 1907, chemists[6] had worked out the decay series
of the uranium isotopes $^{238}U$ and $^{235}U$. And, fortuitously, it turns
out that the different isotopes of uranium degrade to different
isotopes of lead (Pb), with $^{238}U$ decaying to $^{206}Pb$ and $^{235}U$
decaying to $^{207}Pb$. Furthermore, by the 1940s the half-lives of $^{238}U$
and $U^{235}$ (4.47 billion years and 710 million years respectively)
had been worked out. This, in turn, enabled Clair Cameron
Patterson in 1956, working with an iron-nickel meteorite, to
calculate the age of the Earth to be 4.55 billion years[7], an estimate
that is still widely accepted today.

It also turns out that the mineral zircon, a common trace
component of igneous rocks, is perfectly happy to incorporate
either $^{238}U$ or $^{235}U$ into its crystal structure when the mineral is

being formed. But importantly, zircon also actively *excludes* Pb from its structure as the mineral is crystalizing. So, any $^{206}$Pb or $^{207}$Pb found in a zircon crystal can be attributed to the radioactive decay of $^{238}$U and $^{235}$U. It follows that by plotting the ratios of $^{206}$Pb/$^{238}$U versus $^{207}$Pb/ U$^{235}$, a unique date for the age (time since formation) of the zircon crystal can be estimated with pretty fair accuracy (0.1 to 1.0 percent). Because zircons are found in volcanic tuffs, like the rocks of the Mattapan Volcanic Complex, here was a way to finally find out how old they really were.

The first zircon $^{206}$Pb/$^{238}$U-$^{207}$Pb/ U$^{235}$ age dates for volcanic rocks of the Mattapan Volcanic Complex were obtained in 1980 from the old Sally Quarry in Mattapan[8], just a mile or so from The Rock. The results indicated a date of formation of 602 ± 3 million years ago. That, in turn suggested that the Mattapan volcanics were roughly the same age as the rocks from the Avalon Peninsula in Newfoundland, known as the "Avalon Terrane". Those age dates, therefore, indicated that the time of formation of the Mattapan volcanics was related to that of the Avalon Terrane[9]. A later study[10] refined the zircon age dates slightly, indicating ages of between 595 to 597 million years ago.

So now we had a pretty good idea of how old the rocks of the Mattapan Volcanic Complex were. The next question was, where were they formed?

Answering that question took Margaret D. Thompson, a geology professor at Wellesley College near Boston, a good thirty years to work out. Her first paper on the topic was published in 1985 and recognized the presence of ancient volcanic calderas in the Boston area[11]. Much later, her 2007 paper on the Mattapan Volcanic Complex provided measurements of the magnetic declination and inclination of the rocks[12]. When igneous rocks crystalize, whatever magnetic minerals are present line up with the magnetic field of the Earth at the time. Because the direction of the magnetic field indicates where on the globe the rocks were when they crystalized, it's possible to estimate where they were at the time of formation. This is complicated because as rocks may be repeatedly heated up and re-magnetized over time, so you have to be careful that you're seeing the original magnetizing event.

But in any case, Thompson and her colleagues were able to determine that the rocks of the Mattapan Volcanic Complex were most likely located south of the equator when they formed, in a general position south of present-day Africa[12]. But a position of formation north of the equator is also possible and can't be ruled out (Figure 3.4). In either case, over the 600 million years its existence, what would come to be The Rock of Mattapan has moved many thousands of kilometers away from where it was originally formed, eventually welding itself onto what is now North America.

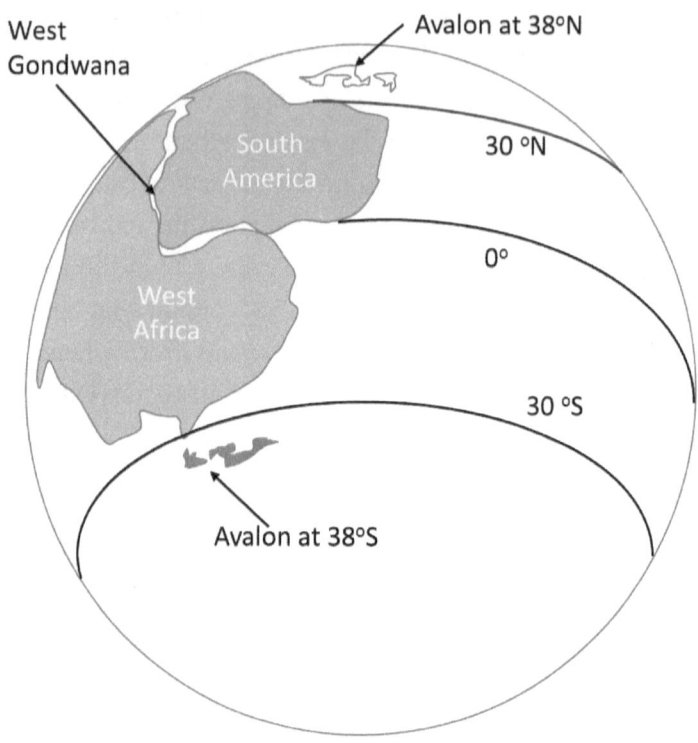

Figure 3.4—Map showing the position of West Gondwana 595 million years ago and the approximate location of the Avalon Terrane relative to what is now the West African and South American continental land masses. The most likely original position of Avalon was at ~38°S, but a position at ~38°N is also possible. Adapted from Thompson et al., 2007[12].

Rocks can mean a lot of different things to different people. For the unfortunate home builder in 1929, The Rock had simply been a pain in the neck, an unfortunate obstacle to his business. To the happy kids who played on The Rock for the next half century, it was a place to get out of doors away from parents and grandparents, and have some fun. By the 1960s, rock removal technology had gotten better, and eventually part of The Rock was removed in order to make room for a nursing home. That nursing home was subsequently transformed into condominiums. It's a pretty safe bet, however, that very few if any of the people who were intimately involved with The Rock over the years—builders, kids, real estate developers—had any inkling of the amazing history that The Rock recorded.

The real miracle of The Rock is that, over more than two centuries, human beings were actually able to piece together its remarkable history. Starting with the Puritans, people immediately took notice of the rocks around Boston. Eventually, people like William Crosby in the nineteenth century began to systematically describe what the rocks looked like and to map where the different rock types were found. After years and years of this, the significance of what those rock types actually meant gradually, very gradually, began to become clear. Finally, the work of Margaret Thompson and her associates, which included radiometric dating and the magnetic record of the rocks, revealed the remarkable journey that The Rock has made and what that journey says about the history of the Earth.

That was no trivial accomplishment.

----------------

As she gazed at the little white rectangular crystals embedded in the broken pieces of The Rock in her hand, the young girl thought how pretty they were. She also noticed how the crystals tended to line up, with their long axes pointed in one direct and the short axes in another. She had never looked closely at any rock before and certainly she'd never noticed how much there was to see in them. As she tossed the rock chips aside before returning to whatever game she and her siblings were playing, she thought to herself that one day she would like to know what those tiny crystals were.

And what they meant.

# REFERENCES

1. Kaye, C.A., 1976. The geology and early history of the Boston area of Massachusetts, a bicentennial approach. U.S. Geological Survey Bulletin No. 1476. US Government Printing Office.

2. Nance, R.D., Murphy, J.B. and Santosh, M., 2014. The supercontinent cycle: a retrospective essay. Gondwana Research, 25(1), pp.4-29.

3. Crosby, W.O., 1880. Contributions to the geology of eastern Massachusetts. Boston society of natural history.

4. Crosby, W.O., 1905. Genetic and structural relations of the igneous rocks of the lower Neponset Valley, Massachusetts. American Geologist, vol. 36, pp. 69-83.

5. Emerson, B.K., 1917. Geology of Massachusetts and Rhode Island. U.S. Geological Survey Bulletin No. 597. US Government Printing Office.

6. Bottwood, B.B., 1907. On the ultimate disintegration products of the radioactive elements. Part II. The disintegration products of uranium. American Journal of Science 23: 77-88.

7. Patterson, C., 1956. Age of meteorites and the earth. Geochimica et Cosmochimica Acta, 10(4), pp.230-237.

8. Kaye, C.A. and Zartman, R.F., 1980. A late Proterozoic to Cambrian age for the stratified rocks of the Boston Basin, Massachusetts, USA. The Caledonides in the USA, Proceedings. Memoirs of the Department of Geological Sciences, Virginia Polytechnic Institute, 2, p.257.

9. Thompson, M.D. and Hermes, O.D., 1990. Ash-flow stratigraphy in the Mattapan Volcanic Complex, greater Boston, Massachusetts. Geological Society of America Special Papers, 245, pp.85-96.

10. Thompson, M.D., Hermes, O.D., Bowring, S.A., Isachsen, C.E., Besancon, J.R. and Kelly, K.L., 1996. Tectonostratigraphic implications of Late Proterozoic U-Pb zircon ages in the Avalon Zone of southeastern New England. Geological Society of America Special Paper 304, pp.179-192.

11. Thompson, M.D., 1985. Evidence for a late Precambrian caldera in Boston, Massachusetts. Geology, 13(9), pp.641-643.

12. Thompson, M.D., Grunow, A.M. and Ramezani, J., 2007. Late Neoproterozoic paleogeography of the Southeastern New England Avalon Zone: insights from U-Pb geochronology and paleomagnetism. Geological Society of America Bulletin, 119(5-6), pp.681-696.

# CHAPTER 4.
## GREEN IS FOR COPPER

It all started with a comic book and a chemistry set.

The father of the young boy from West Point had been reassigned to an overseas posting, as regularly happened in the Army. This posting, beginning in the summer of 1960, was to a military hospital located near the Saarland region of southwestern Germany. 1960 was the height of the Cold War, and the U.S. Army was deployed in Germany to discourage the Soviet Union from attempting an invasion of Western Europe. If such a war actually broke out, it would be critical to have medical facilities already in place to handle what were certain to be very high casualties. One of these facilities was the 98th General Hospital at Neubrücke, which was located in the valley of the Nahe River. The young boy's father's new job was to be the Executive Officer of the Hospital.

The boy, now eight years old, was an indifferent student in school but he loved to read comic books. One comic book in particular, which was devoted to science fiction, had recently grabbed his attention. The story revolved around a group of people who had been sent by spaceship to a distant planet in order to build a colony. The planet was much like Earth, and it had fertile soils which yielded abundant crops for the new settlement. However, after a year or so, a problem developed. People began to exhibit symptoms of mental illness which included hallucinations combined with feelings of paranoia. At first, it was thought that these symptoms were simply a reflection of the colony's isolation from the rest of humanity. But one person in the group thought something else, something much more serious, was going on. Specifically, this person had a chemistry set designed to analyze soils. And, after much sampling and analysis, he determined that the soil contained a toxic compound that was being taken up by the crops and transmitted to the people from the food they were eating.

Long story short, the soil-analyzing chemist identified the toxic compound, figured out how to keep it out of the food supply, and the colony was saved. That sounded cool to the young boy, who resolved to get his own chemistry set and go around testing the soils of Neubrücke for various imagined toxic compounds.

The idea of their son, who was prone to getting into trouble (he once set a fire in West Point and ended up burning five acres of woods), mixing chemicals together in the basement of their house didn't appeal to his parents. But the boy was persistent, and after much badgering, his parents finally relented and got him a chemistry set as a Christmas present. Armed with the chemistry set, the boy immediately set about analyzing the local soils for the presence or absence of various chemical compounds.

The chemistry sets that you can buy today in toy stores have removed any chemicals that can even remotely be considered dangerous. Too much liability for the toymakers. But in 1960, chemistry sets were actually tools that enabled the user to do real, if unsophisticated, chemical analyses. This particular chemistry set included assays for some of the more common metals found in sediments and soils, including iron and copper. The boy excitedly went outside and scooped up a cupful of soil. The first assay he tried was for the presence or absence of iron, which unsurprisingly indicated that, yes, iron was indeed present. Next, the boy tried the assay for copper, fully expecting to find nothing. The procedure in the instructions required putting a few grams of soil in a beaker, adding several milliliters of hydrochloric acid (HCl) (which you'll never find in a modern toy chemistry set) and adding an indicator reagent. Copper ions ($Cu^{2+}$) have the interesting property that, when surrounded by chloride ($Cl^-$) ions (supplied by the hydrochloric acid), they turn a bright green color. Sure enough, the soil-water mixture in the boy's test tube turned a bright green.

Copper!

The boy vaguely knew that copper could be toxic, and he excitedly ran to tell his father. His father, however, was unimpressed. Copper in small quantities is actually an essential nutrient for plants and animals, and so its presence in a soil didn't necessarily indicate toxicity. Furthermore, the local soils had been farmed for centuries without any noticeable problems. So, while the presence of copper was interesting and perhaps unusual, it certainly wasn't something to worry about. Disappointed, the boy returned to his chemistry set to search for other (hopefully) toxic chemicals.

Very much in passing, he wondered where the copper in the Neubrücke soil might have come from.

-------------------------

It turns out that copper is central to the history of this part of Germany. The local Celtic tribes had been mining copper and zinc for hundreds of years prior to the arrival of the Romans in the first century BC. In fact, it is the copper and lead ores, as well as gemstones such as agate also found in in the vicinity of the Nahe River[1] that drew the Romans to this region in the first place. The Romans were not in the habit of conquering people arbitrarily. Rather, they were interested in exploiting (that is, stealing) the local resources for themselves. If an unconquered area had valuable resources, such as metal ores, the self-interested Romans would go to the trouble and expense of adding it to their empire.

The conquest of Germania was undertaken in in 12 BC by Augustus Caesar's generals Germanicus and Tiberius. Tiberius, partly because of this successful military campaign, later succeeded Augustus Caesar as the second Roman Emperor. By 6 AD, Rome controlled much of southwestern Germany as far as the Elba River, including what is now the Saarland.

As the Romans took over the Nahe Valley region, they immediately began to apply their relatively sophisticated mining technology. Soon the mines were producing significant amounts of copper, zinc, lead, iron, and even a smattering of gold and silver. The other thing that attracted the Romans to this part of Germany was the presence of hot mineral springs which were ideal for the bathhouses central to Roman social life. In German, the noun "Bad" means "bath" or "spa", and beginning with the Roman occupation, towns with names like Bad Sobernheim, Bad Münster, and Bad Kreuznach grew up around the mineral springs of the Nahe River Valley.

As the Romans tightened their grip on Germania, they built camps for two Roman legions at the confluence of the Main and Rhine Rivers which is now the city of Mainz. Mainz is about a hundred kilometers from the mineral districts of the Nahe Valley, and it soon developed an important metal-working industry smelting lead, zinc, and copper[2]. There is not a lot of archaeological evidence as to which metal ore deposits were mined by the Romans, but isotopes of both lead and copper clearly show that bronze, brass, and lead objects found at various sites throughout Europe were made from ores mined in the vicinity of the Nahe Valley[2].

These days the best-known Nahe Valley copper mine, called the Hosenberg mine, is located near the town of Fischbach. The mine has been worked since at least the year 1400 AD, and by the 16th century there were about 300 miners working the mine. Over the centuries the miners excavated huge vaulted rooms in the surrounding country rock in order to extract the copper ores that included native copper, bornite, chalcocite, and malachite. Production at the mine ceased in 1825, but in 1975 the mine reopened as a tourist attraction to show off the spectacular interior of the mine. Just how that copper mineralization was emplaced in the Nahe Valley reflects a long complex geologic history, and one that illustrates the kinds of events that have to happen in order to produce economically important metal ores.

------------------------

Five hundred million years ago, land masses that are now the continents of Africa, North America, and Eurasia began to converge with Gondwana (Fig. 3.1), eventually forming the supercontinent Pangea. The term "converge" doesn't quite capture the flavor of the violence that this entailed. For the next 250 million years, the boundaries of various plates alternately crashed together forming volcanos and mountains, or split apart to form basins. Those basins created seas that subsequently trapped sediments eroding from the mountains.[3,4] All of this activity went on until the Atlantic Ocean began to open up in earnest 130 million years ago. The geologic remnants of these numerous mountain-building and basin-forming events are known today as the Variscan Belt which covers much of Europe (Fig. 3.1). As you might expect, remnants of the Variscan Belt are also found in Africa and North America. In North America, rocks associated with the Veriscan events formed the Appalachian Mountains in Pennsylvania and New York, and the Ouachita Mountains in Arkansas and Oklahoma [3]. In addition, these collisions also served to weld portions of Avalonia (Chapter 2) onto both North America and Eurasia.

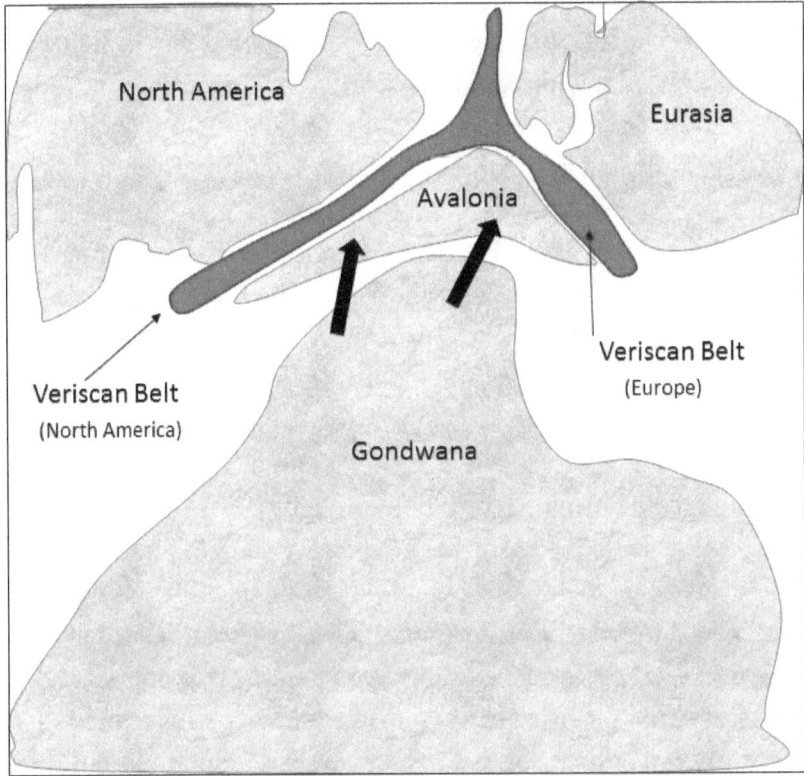

Figure 3.1-- Plate collisions recorded by the Veriscan Belt between 380 and 280 million years ago that contributed to the copper and lead mineralization now found in the Nahe Valley, Germany.

The violent plate collisions that formed the Variscan belt are reflected both in their structure and chemistry. As the plates periodically ground together and then pulled apart, numerous structural basins were formed. One of these is the Saar-Nahe basin of Germany[5]. The volcanic rocks underlying the basin include the dark basalts that indicate mantle-derived lavas bubbling up as the tectonic plates were wrenched apart, and andesites and rhyolites formed by partial melting of sediments. Neubrücke, located in the valley of the Nahe River, is surrounded by rhyolites that form steep ridges, and basalts that form more level terrain.

As you might expect, the steep rhyolite ridges are not

particularly promising terrain for agriculture. However, when the Romans occupied the region in AD 6, and being unwilling to do without wine, they introduced the cultivation of grapes. Over the centuries, it was discovered that the poor, well-drained soils underlying the steep rhyolite ridges were actually ideal for the cultivation of wine grapes. Today, the steep ridges surrounding the Nahe and Mosel valleys have been carefully terraced and support a thriving wine industry. The wine fests, which are held in the fall of every year in the Nahe and Mosel regions, celebrate each year's grape harvest and serve as an opportunity to sample the new wines and last year's wines as well. That, incidentally, is where the young boy got his first taste of wine. Germans aren't as fussy about underage drinking as are Americans.

But while the basin-forming and mountain building in the Variscan orogenic belt can explain the presence of rhyolites and basalts and the resulting steep topography surrounding the Nahe and Mosel valleys, it doesn't quite explain the rich copper and lead mineralization that the Celts, Romans, and Germans exploited so vigorously over the years. The last phases of mountain-building events that led to the Versican belt are often characterized by meteoric or sea waters percolating deep into the Earth. As the waters are heated up by the magma masses at depth, they strip the surrounding rocks and sediments of soluble metals such as copper, lead, gold and silver. These now metal- and sulfur-laden waters are then forced upward due to the expansion of the water, cooled, and the metals drop out of solution as sulfide minerals. This has the effect of greatly concentrating the metals into rich ores that are much easier to mine. And the soils that develop on rocks associated with these ore deposits often contain relatively high concentrations of metals such as copper, which explains the copper found in the soils of Neubrücke by the young boy with his chemistry set.

------------------------

Humans, like all living creatures, make a living by utilizing whatever natural resources happen to be available to them. First, the Celts learned how to smelt copper and zinc sulfide rocks, probably in pottery kilns, in order to produce brass. The Romans, seeing the economic opportunity presented by the copper and lead deposits, moved in, conquered the local tribes and coopted the metal-producing industry for themselves. After the Roman Empire collapsed, the Germans continued the mining industry which eventually contributed to the industrial revolution in Europe, and which made Germany one of the most powerful industrial nations in world.

Incidental to all this brave history, it also gave one young boy with a chemistry set pause to wonder just where the copper in the German soil might have come from.

## REFERENCES

1. Götze, J., Tichomirowa, M., Fuchs, H., Pilot, J. and Sharp, Z.D., 2001. Geochemistry of agates: a trace element and stable isotope study. Chemical Geology, 175(3), pp.523-541.
2. Durali-Müeller, Soodabeh. 2005. Roman lead and copper mining in Germany their origin and development through time, deduced from lead and copper isotope provenance studies. Dissertation zur Erlangen des Doktorgrades Der Naturwessenschaftern, Johann Wolfgang Goethe-Universität in The young manfurt am Main.
3. Matte, P., 1986. Tectonics and plate tectonics model for the Variscan belt of Europe. Tectonophysics, 126(2), pp.329-374.
4. Matte, P., 1991. Accretionary history and crustal evolution of the Variscan belt in Western Europe. Tectonophysics, 196(3), pp.309-337.
5. Schmidberger, S.S. and Hegner, E., 1999. Geochemistry and isotope systematics of calc-alkaline volcanic rocks from the Saar-Nahe basin (SW Germany)–implications for Late-Variscan orogenic development. Contributions to Mineralogy and Petrology,

135(4), pp.373-385.

# CHAPTER 5.
## MAKING ROCK CREEK PARK

When the young girl's family moved to Washington D.C, they rented a house in Bethesda, Maryland, which in those days was just a little town. In the 1960s, many of the houses in Bethesda were owned by military or diplomatic families. These people were routinely being posted overseas for one or more years, and so they would often rent their houses out until they rotated home again.

The house the young girl's family rented from one of these families was nice, but it did not have air conditioning. That's a problem in Washington D.C. where the summers are brutally hot and humid. That being the case, the family was always looking for ways to keep the kids cool during the long, hot summer days.

One good way to keep cool was to go to Rock Creek Park, which was only a couple of miles from their rented house. Rock Creek is deeply incised into the local piedmont rocks and forms a rugged valley, which as the name suggests, is filled with rocks and boulders (Fig. 5.1). But also, the steep valley sides surrounding the rock-filled creek bed are covered with huge shade trees. That shade, and the cold water running in the creek, made it one of the few places in the city that was always cool, even on the hottest days.

Figure 5.1. Rock Creek. National Park Service file photo.

One day, the young girl's family was having a picnic in Rock Creek Park with their friends the Brannigans. Anne Brannigan was the same age as the young girl, about eight years old, and soon the two friends and Anne's older brother were happily wading and splashing in the cool water of Rock Creek. But it had rained recently, and the water in the creek was relatively high and fast. As it happens, the young girl waded too far out in the stream, slipped on a rock, and before she knew it she was being washed downstream. Anne's brother, who was thirteen years old, saw what had happened. He quickly waded out to where the young girl was struggling to keep her head above water, grabbed her, and dragged her out of the creek. She was dazed and frightened, but otherwise unhurt.

No longer interested in wading in the creek, the young girl sat on the stream bank and idly began to regard the huge boulders that surrounded her. Some of the rocks, she noticed, had elliptical blobs embedded in them that were lighter in color than the surrounding black rock. Furthermore, she noticed that some of blobs were elliptical in shape and seemed to line up parallel to each other. That was odd, she thought, why would that be? Those "blobs", incidentally, are still clearly visible in Rock Creek Park today (Figure 5.2).

But the young girl mostly remembers that the rocks looked old.

As old as the hills.

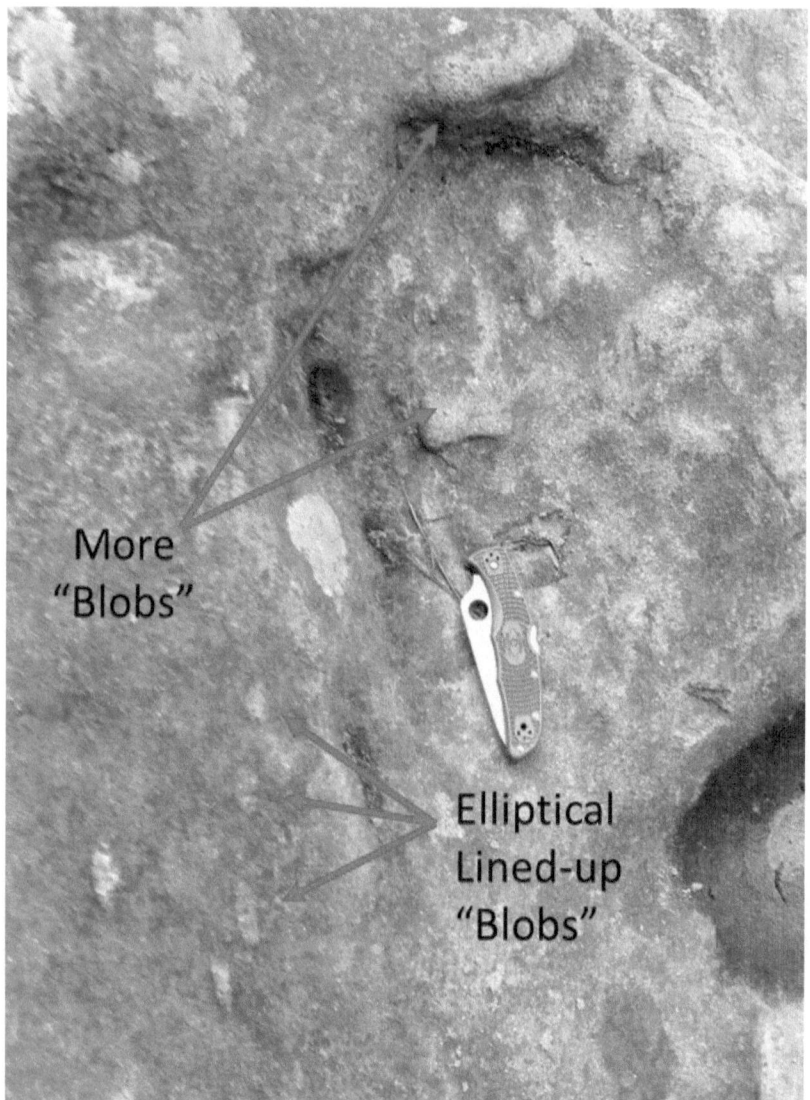

Figure 5.2—"Blobs" of rock inclusions visible in the rocks of Rock Creek Park. Pocket knife is for scale.

------------------------------

The city of Washington D.C. is remarkable in a lot of ways that include stately buildings and monuments, a deep and consequential history, and bewildering roads that sometimes go around in circles. But the actual landscape of most of the city is decidedly unremarkable. The older part of the city, which includes the Capital, the White House, and the Washington Monument was built on drab, flat coastal plain sediments. Thus, shining spectacular vistas like what you can see in Denver or Salt Lake City, are largely absent in Washington D.C.

Rock Creek Park is the sole exception to that rule. From the earliest founding of Washington D.C., Rock Creek not only provided welcome relief from the summertime heat and humidity, it also provided a remarkably beautiful landscape. That landscape was so lovely and so strikingly different from the rest of the city that it was set aside as a national park in 1890. Today, commuters will take a winding two-lane road named Beach Drive that leads through Rock Creek Park to downtown Washington, not because it's particularly fast or convenient (it's not), but simply because it's such a pretty drive. That being the case, it's reasonable to wonder just why Rock Creek's landscape is so different and so much more beautiful than its humdrum surroundings.

One obvious difference between Rock Creek Park and the rest of Washington is that it is a deep valley carved into the hard, crystalline rocks of the piedmont. From the beginning, however, the rocks that made up the piedmont were enigmatic. In some places the rocks were clearly of igneous origin such as granites, basalts, and gabbros (basalts that cooled and crystalized before reaching land surface). In other places the rocks were more foliated and consisted of metamorphic schists and gneisses that seemed to be derived from sedimentary rocks. But where the rocks had come from or what had caused their puzzling variety was simply a mystery. It was a mystery that took more than a hundred years to solve.

The first geologic map of Washington D.C., which included Rock Creek Park, was published in 1901 by two U.S. Geological Survey (USGS) geologists named Arthur Keith and Nelson Horatio Darton (Figure 5.3A)[1]. The map, which clearly was intended as a preliminary reconnaissance, shows most of Rock Creek running through a rock type that Keith and Darton described as a "granite gneiss". Just west of the creek, and standing topographically higher is a rock type they described as a "biotite granite". Importantly, Keith and Darton mapped the contact between the two rock types as a sharp boundary, implying that some sort of discontinuity separated them. But what kind of discontinuity?

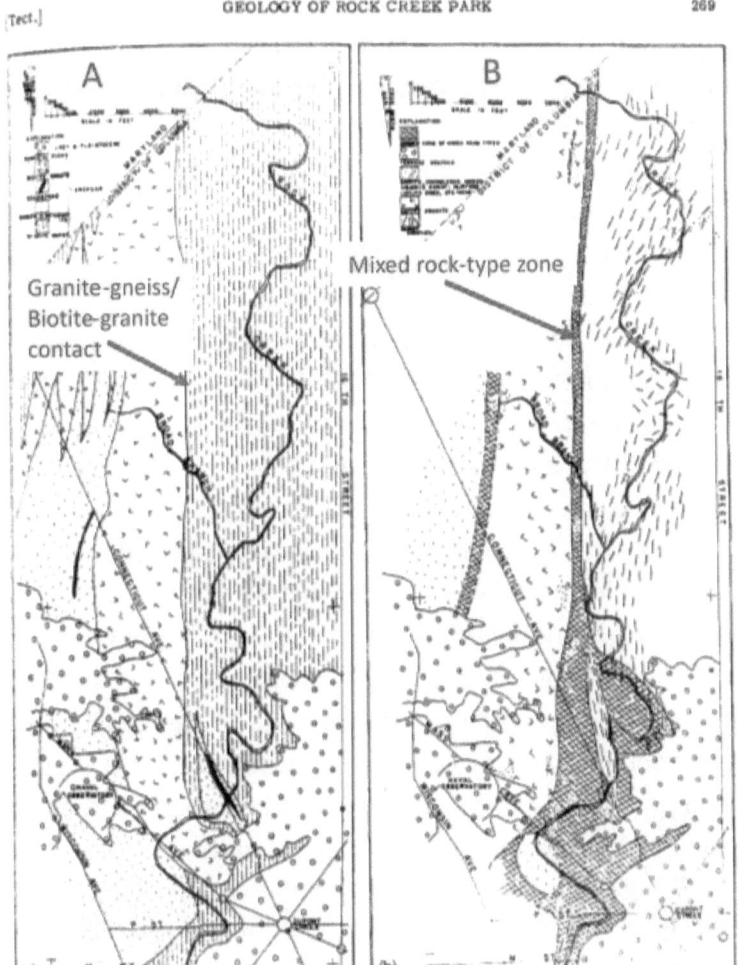

Figure 5.3—A comparison of (A) the Keith and Darton map of 1901 and (B) the Robert Fellows map of 1950 of Rock Creek Park. Note the similarity of Keith and Darton's boundary between the different granites to the west of Rock Creek and Fellow's "mixed rock type zone".

Years later, in 1950, another USGS geologist named Robert Fellows also published a geologic description of Rock Creek Park[2]. Fellows was an interesting person. He was originally assigned to

47

the Alaskan Section of USGS, and between 1947 and 1949 had been transferred to Washington D.C. in order to assume an administrative position. Apparently he found administrative work less than compelling, and so together with his wife he began doing fieldwork in Rock Creek Park, as he later explained "on weekends, holidays, and summer evenings".

But what he saw puzzled him. He could see immediately that the Keith and Darton map lacked a good deal of detail about rock textures that he could plainly see, and he seems to have resolved to improve their map. Also, Fellows noted that what Keith and Darton were calling a "granite gneiss" (where the streambed of Rock Creek was located (Fig. 5.3A) actually looked more like a "schist complex". In other words, the rocks underlying Rock Creek were actually highly metamorphosed sediments, not a true granite. Fellows also could see that what Keith and Darton were calling a "biotite granite" west of the creek was in fact a true granite, but it was shot through with quartz veins and aplite (fine-grained rocks of igneous origin) dikes. All of which indicated a long and complex history of sediment deposition, extreme heating and compression (forming the schists and gneisses), and igneous intrusion by both basaltic and granitic magmas.

What was particularly struck Fellows, however, was that what Keith and Darton had mapped as a simple sharp contact between the biotite granite and the schist (Fig. 5.3A), seemed to be more of a "zone of mixed-type" rocks (Fig. 5.3B) that:

> *Show marked effects of intrusive activity in the form of aplite dikes, amphibolite dikes, heavy concentrations of quartz, xenoliths, bodies of soapstone, extreme contortion, and slickensiding[2].*

That quote is interesting for a couple of reasons. For one thing, Fellows noted the presence of *xenoliths* imbedded in the rocks. Xenoliths are pieces of rock that have a different origin from the surrounding country rock. In sedimentary rocks, and metamorphic rocks derived from sedimentary rocks, xenoliths can be formed by submarine avalanches careening down the sides of oceanic trenches before coming to rest in more fine-grained sediments. When the sediments are metamorphosed later in schists or gneisses, the rock chunks can be preserved as xenoliths. In igneous rocks, xenoliths can be picked up by liquid magmas intruding violently into preexisting country rocks. If the magma cools before the country-rock chunks melt, they too can be preserved as xenoliths. Xenoliths, either of the sedimentary/metamorphic variety[3], or the igneous variety are one of the defining features of the rocks found in Rock Creek today (Figure 5.4)

Figure 5.4—A xenolith embedded in a granitic matrix in Rock Creek Park. Note how the texture of the rock fragment differs from the surrounding matrix. This kind of rock may be what Robert Fellows referred to in 1950 as the "mixed rock-type zone". Pocket knife is for scale.

The other thing that Fellows noticed was that mixed rock-type zone showed evidence of *extreme contortion, and slickensiding.* The "extreme contortion" was evident from the visible folds and bends in the rock which certainly reflected compressive forces and could indicate active faulting. Similarly, "slickensides" are caused by frictional movement between rocks along the two sides of a fault. Could it be that the "mixed rock-type zone" indicated the presence of one or more faults? Fellows clearly thought that his observations deserved to be looked into in more detail. But unfortunately, he died suddenly before his 1950 paper was published, and so that task would have to be taken up by others.

The next person to take a close look at the geology of Washington D.C. was another USGS geologist named Paul M. Johnson in 1964[4]. By that time, the piedmont rocks underlying Washington D.C. had been examined and mapped by several geologists, including Ernst Cloos of Johns Hopkins University in Baltimore[5], but if anything the rocks remained as enigmatic as ever. Paul Johnston, like Fellows before him, noted the odd-looking "mixed rock-type zone" and the puzzling association of basaltic and granitic igneous rocks that had intruded into metamorphosed sediments, and bravely went on to speculate how the rocks might possibly have formed. He wrote[4]:

> The history of these rocks begins with the deposition
> of sediments in the sea at least 440 million years
> ago. After consolidation, the rocks were raised
> above the sea and intruded by mafic magmas, which
> in some places reached the surface and resulted in
> volcanic activity. Much later, in late Paleozoic
> time, strong compression forces, acting in a
> northwest or southeast direction buckled the earth's
> crust and compressed the beds into tight folds.

In retrospect, Johnson's suggested history of Rock Creek's piedmont rocks turned out to be fairly accurate. At that time (1962), the idea that ocean basins could open and close over time was a new concept[6]. Furthermore, that ocean basins opening and closing could cause continents to collide, something referred to today as a Wilson Cycle[7], was also a new concept. And, as

Johnson suspected, that general sequence of events is pretty close what led to the formation of the piedmont rocks of Washington D.C. and to Rock Creek Park itself.

    A possible sequence of events that formed the rocks in Rock Creek Park goes something like this[8]. Sometime about 530 million years ago (Figure 5.5A), the proto-North American continent was separated from the supercontinent Gondwana and was surrounded by a vast shallow sea teaming with newly-evolved invertebrate organisms. However, as Gondwana edged closer to North America, the tectonic plates carrying the continents began to collide causing a series of volcanic islands to develop offshore. Such volcanic islands tend to erode rather quickly, dumping huge amounts of sediment into the surrounding seas. So, the first part of Paul Johnson's hypothesized sequence, *"The history of these rocks begins with the deposition of sediments in the sea at least 440 million years ago",* looks pretty good.

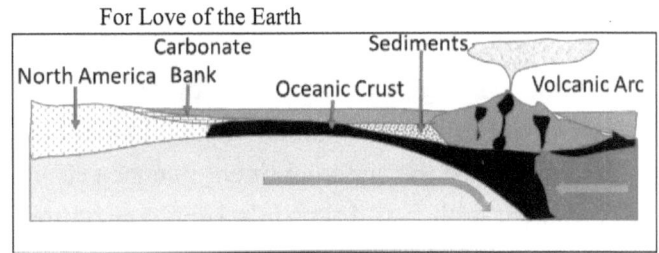

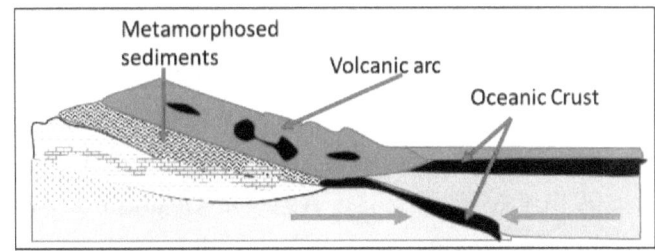

Figure 5.5—Some of the events that produced the piedmont rocks underlying Rock Creek Park we see today. Adapted from U.S. Geological Survey, 2017[8].

By 440 million years ago (Figure 5.5B), the collision between North America and the volcanic rocks of the island arc and their associated sediments was well under way. This caused a massive mountain-building episode on the margin of North America known as the Taconic Orogeny, named after the Taconic Mountains of New York. These must have once been a spectacular mountain range, probably comparable to the modern Himalayas. But in any case, that is basically what Paul Johnson had envisaged when he wrote *after consolidation, the rocks were raised above the sea and intruded by mafic magmas, which in some places reached the surface and resulted in volcanic activity.*

But the collisions caused by the shrinking ocean basin east of North America were just the beginning. By 370 million years ago, another series of volcanic arc islands, known as the Avalon Terrane (Chapter 2), collided with North America causing another mountain-building episode known as the Acadian Orogeny. Finally, Gondwana collided with North America around 320 million years ago causing yet another mountain-building episode known as the Alleghenian Orogeny. That marked the final assemblage of the supercontinent Pangaea (Chapter 4), also largely consistent with Paul Johnson's 1964 conjectures when he wrote *much later, in late Paleozoic time, strong compression forces acting in a northwest or southeast direction buckled the earth's crust and compressed the beds into tight folds.*

While Paul Johnson's suggested picture of events is reasonable based on what we know now, the fieldwork that nailed down the details of that story was carried out between 1977 and 1991 by A.H. Fleming, A.A. Drake, and Lucy McCarten of USGS. This painstaking fieldwork, which took years to complete, culminated with a new geologic map of Washington D.C. that was published in 1994[9]. That map went a long way toward clearing up some of the mysteries surrounding the origin of the piedmont rocks. But more relevant to our story, those studies also showed just what had formed the steep-sided valley that now holds Rock Creek Park in the first place.

As it happens, land masses don't always collide head on. In many places, the San Andreas Fault of California being a modern example, these rock masses simply slide past each other from left to right (or right to left). These so-called "sheer zones" have been found in mountain terrains throughout the world. It wasn't until 1994, however, that it was recognized that the valley holding Rock Creek Park is one of those sheer zones[10]. That, in turn, explains the sharp boundary between rock types mapped by Keith and Darton (Fig. 5.2A), and the "zone of mixed rock-type" mapped by Robert Fellows (Fig. 5.2B). But what's even more intriguing, the Rock Creek Sheer Zone is not your ordinary every-day sort of sheer zone. But rather, it's a sheer zone with a very unusual history.

During the Taconic Orogeny (Figure 5.5B), a sheer zone developed in the metamorphic rocks at the base of the growing mountain chain. The movement along that sheer zone was in a left-to-right direction, occurred over tens of millions of years and had the effect of grinding up, folding, and foliating the rocks. After the Taconic event, the sheering movement ceased and for the next hundred million years or so, everything was quiet. But during the final stages of the Alleghenian Orogeny, movement along the sheer zone started up again. This time, however, instead of a left-to-right movement, the movement was from right to left. This, of course, caused further grinding, faulting, and deformation of the rocks.

The 300 million years that followed the Alleghenian orogeny saw the erosion of the various mountain chains that had been built by the colliding continents. Much of the metamorphism that heated and squeezed the piedmont rocks underlying Washington D.C. occurred tens of kilometers deep within the earth. But as erosion stripped off those mountains, depositing much of the sediment in what is now the eastern coastal plain of the United States, the piedmont we see represents only the roots of those ancient mountains. It wasn't until the Rock Creek Sheer Zone had been exposed by erosion that the last phase of making Rock Creek Park could commence.

Thanks to the heat and compression that the piedmont rocks have been exposed to over the millennia, they are highly crystallized, extremely tough, and notably resistant to erosion. The exception is the Rock Creek Sheer Zone. Because of the grinding, faulting, and slippage along the zone, its rocks are more shattered and thus more susceptible to erosion than the surrounding rocks. So, as erosion uncovered the Rock Creek Sheer Zone, it became a natural conduit for water to run off the land, and a creek bed developed. Furthermore, and again because of the relative weakness of the rocks, it was natural for the creek bed to erode more deeply into the sheer zone. That erosion is what led to the steep-sided valley that now holds Rock Creek (Figure 5.6), and that's one reason why the valley is so strikingly beautiful.

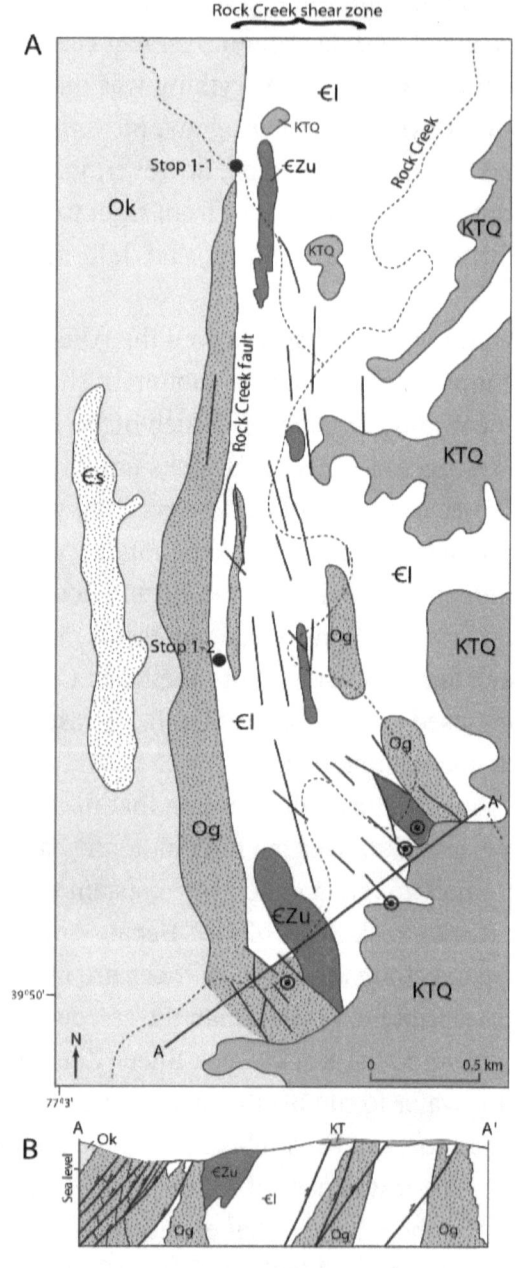

Figure 5.6—Geologic map of Rock Creek Park as mapped by Avery, Drake, and McCartan between 1977 and 1992[9].

So, Rock Creek Park was a long time in the making. What we see today is the end result of more than 500 million years of continents being split apart, ocean basins growing, those ocean basins reversing and closing, continents colliding, and massive mountain ranges being lifted out of the sea. If one Wilson Cycle consists of a continent splitting apart, forming an ocean basin, and then closing back up again to build massive mountain chains, then the rocks of Rock Creek Park record not one, but as many as three Wilson Cycles[11].

As the young girl had guessed, while she was sitting on the bank of Rock Creek shivering from her plunge in the cold water and wondering about the rocks she saw, they are indeed as old as the hills.

-------------------------------

At about the same time the young girl and her family were picnicking in Rock Creek Park, the young boy was happily jumping from rock to rock along a creek known as Spout Run on the other side of the Potomac River in Virginia. His father had just been reassigned from Germany to the Army Research and Development Command in Washington, and he had bought a house on a hill overlooking Spout Run. Like Rock Creek, it's a fast-moving stream in a valley littered with boulders. And like Rock Creek, it owes its existence to the sheering and grinding movements of faults related to the Rock Creek Sheer Zone. But unlike the young girl, the boy never noticed the blobs of quartz, other rock fragments, and xenoliths embedded in the rocks. So, during the 1960s, our young girl and boy were living just a few miles from each other.

They wouldn't meet for another ten years.

## REFERENCES

1. Darton, N.H. and Keith, A., 1901. Washington, DC-Md.-Va. US Geological Survey Geological Atlas of the United States,

Folio, 70.

2. Fellows, R.E., 1950. Notes on the geology of Rock Creek Park, District Of Columbia. Eos, Transactions American Geophysical Union, 31(2), pp.267-277.

3. Pavlides, L., 1989. Early Paleozoic composite mélange terrane, central Appalachian Piedmont, Virginia and Maryland; Its origin and tectonic history. Geological Society of America Special Papers, 228, pp.135-194.

4. Johnston, P.M., 1964. Geology and ground-water resources of Washington, D.C., and vicinity. U.S. Geological Survey Water Supply Paper 1776, US Government Printing Office, Washington, D.C., 97 pp.

5. Cloos, E. and Cooke, C.W., 1953. Geologic map of Montgomery County and the District of Columbia, scale, 1:62,500. Maryland Department of Geology, Mines, and Water Resources, Baltimore. Maryland.

6. Hess, H.H., 1962. History of Ocean Basins. In A. E. J. Engel, Harold L. James, B. F. Leonard (eds.), Petrologic Studies: A Volume in Honor of A.F. Buddington. Geological Society of America. pp. 599–620, Boulder, CO.

7. Wilson, J.T., 1966. Did the Atlantic close and then re-open? Nature vol. 211, pp. 676-681.

8. U.S. Geological Survey, 2017. Geology of National Parks,

9. Fleming, A.H., Drake, A.A. and McCartan, L., 1994. Geologic map of the Washington West quadrangle, District of Columbia, Montgomery and Prince Georges counties, Maryland, and Arlington and Fairfax counties, Virginia (No. 1748).

10. Fleming, A.H. and Drake Jr, A.A., 1998. Structure, age, and tectonic setting of a multiply reactivated shear zone in the Piedmont in Washington, DC, and vicinity. Southeastern Geology, 37(3), pp.115-140.

11. Southworth, S., Drake, A.A., Brezinski, D.K., Wintsch, R.P., Kunk, M.J., Aleinikoff, J.N., Naeser, C.W. and Naeser, N.D., 2006. Central Appalachian Piedmont and Blue Ridge tectonic

transect, Potomac River corridor. *in* Pazzaglia, F.J., ed., Excursions in Geology and History: Field Trips in the Middle Atlantic States: Geological Society of America Field Guide 8, pp.135-167.

# CHAPTER 6.
# THE ROAD TO GETTYSBURG

One Saturday morning in July, the young girl's father announced that the family was going to take a day trip to visit the famous battlefield at Gettysburg, Pennsylvania. Gettysburg was only about a three-hour drive from their home in Bethesda, and the father thought that a visit to the site of the decisive battle of the American Civil War would "be educational" for the kids.

The young girl and her siblings could only groan. Their father, who was genuinely fond of his four children, was also a classic workaholic. He had served in the Army during World War II, but had never been posted overseas. The Army, needing engineers, had sent him to engineering school. But the war had ended before he finished his training and the Army promptly discharged him. He finished college, majoring in business administration at Boston College, married in 1952, and took a job with an insurance company in Chicago. For the first five or six years of his career, he traveled five days a week selling insurance in Illinois. The kids' earliest recollection of their father was that he was seldom home.

In 1960, when the young girl was six years old, the father was transferred to Washington D.C. where, because of his background in both civil engineering and business, he was assigned to writing performance bonds for construction projects. In the early 1960s, construction in Washington D.C. and the surrounding suburbs was booming and underwriting construction bonds was big business for insurance companies. There was a lot of work to do, and the young girl's father, with four kids to support, was willing to do it. He never worked less than six days a week and he often went into the office on Sunday afternoons as well. Periodically, he would start feeling guilty about not spending enough time with his children, and on those occasions would load everybody into the family station wagon and go for a day trip somewhere. Today, when the young girl was eleven years old, the destination was Gettysburg, Pennsylvania.

It was July in Washington D.C., and so the weather was stiflingly hot and humid, the station wagon was not air-conditioned, and the three-hour drive to Gettysburg was three hours of misery for the kids. Nor did the misery stop when they got to Gettysburg. They spent some time viewing the famous cyclorama, painted in 1882 by the French artist Paul Philippoteaux. That was good because it was both entertaining and air conditioned. But the father soon shuffled the kids back to the hot station wagon, and he commenced driving around the battlefield in the oppressive heat.

--------------------------------

The Battle of Gettysburg is famous for many reasons. But it is particularly remembered because it represented "the high water mark" of the Confederacy, when Longstreet's Assault (also known as Picket's Charge) came within a whisker of winning the battle and with it the Civil War. Instead, the Union line held, the Confederates were forced to leave the field having suffered dreadful casualties, and thereafter the war turned firmly in favor of the Union. But the Battle of Gettysburg is also famous for how geology affected the course of the battle.[1]

In early 1863, General Robert E. Lee and his Army of Northern Virginia was on a winning streak. In May of that year, Union General Joseph Hooker engineered a surprise and well-executed crossing of the Rappahannock River with his Army of the Potomac, and immediately threatened Lee's exposed flank near the town of Chancellorsville. But rather than retreat from the much larger Union Army, Lee divided his forces and sent General Stonewall Jackson and 30,000 men around the Union right flank. The sheer audacity of that move was so extreme that Hooker refused to believe the reports of his own scouts who tried to tell him what the Confederates were doing. Jackson's assault crushed the exposed Union flank and sent the Army of the Potomac into a frenzied retreat. The Battle of Chancellorsville is widely considered to be Lee's crowning masterpiece of generalship. Erwin Rommel, the best German General of World War II, traveled to Chancellorsville in the 1930s in order to study the battle personally.

So Lee, tired of constantly being on the defense in Virginia and hoping to attract help from England, decided to invade Pennsylvania. The army that Lee led was supremely confident, and the soldiers had as solid a faith in their commanding general as any veteran army that has ever marched. Characteristically, Lee's move north took General Hooker completely by surprise, leading President Lincoln to replace him with General George Meade. This change of command rattled the Army of the Potomac, which was bad enough, but it was confounded by the fact that Meade had no idea where Lee was or where he was going. After more than a week of uncertainty, a detachment of Union cavalry stumbled onto a column of Confederate infantry near the little town of Gettysburg. Legend has it that the Confederates, many of whom were barefoot, had heard that there was a shoe factory in Gettysburg, and the only reason they went there at all was to see if they could find some boots.

The Union cavalry commander immediately perceived that Seminary Ridge, which protruded above the surrounding flat farmland west of Gettysburg, would be strong defensive position. So he dismounted his cavalrymen, placed them and their new-fangled lever-action repeating rifles on Seminary Ridge facing the Confederate infantry, and sent for help. The next morning, the Confederate infantry attacked the dismounted cavalrymen and the Battle of Gettysburg had begun. Things initially when well for the Union, as two full corps of Union infantry soon reinforced the dismounted and outnumbered cavalry. But then two corps of Confederate infantry, who simply had orders to come to Gettysburg, surprised the Union army's right flank, and just like at Chancellorsville, the Yankees had no choice but to turn and run for it.

But fortunately for the Yankees, there were another series of hills and ridges just east of Gettysburg, and the troops were able to reform their line of defense on the high ground of Culp's Hill and Cemetery Ridge as night was falling on the first day. For the next two days, the battle consisted entirely of the Rebels trying to drive the Yankees off of those hills and ridges. On the second day, Rebels commanded by General James Longstreet tried to move around to the left of the Yankee position. This led to desperate fighting over flat farmland in front of the ridges, particularly in a wheat field and peach orchard, and in a pile of curious black rocks known locally as the Devil's Den. Longstreet's advance was finally stopped by a spirited Yankee defense of another black stone-covered hill known as Little Round Top.

On the third day (July 4th), the Rebels mounted the now-famous Longstreet's Assault, trying to dislodge the Yankees from Cemetery Ridge in the center of the Yankee line. When that assault failed, the Confederates had no choice but to withdraw. For the first time, the Army of the Potomac had decisively won a battle against Lee's Army of Northern Virginia. On that same day, General Ulysses S. Grant captured the town of Vicksburg, Mississippi giving the Union control of the entire Mississippi River. After these twin defeats in the summer of 1863, the South effectively had no chance to win the war.

The location of Seminary Ridge on the first day, the flat farmlands with its wheat fields and peach orchards, Culp's Hill, Cemetery Hill, and Little Round Top, and the curious "black rocks" that covered those hills, dictated virtually every aspect of the battle. It's natural, therefore, to wonder just why the ridges, the hills, the black rocks, and the flat farmland were there in the first place. The answer to that question, which ironically mirrors why there was a Battle of Gettysburg in the first place, was all about the violence that can accompany rifts and separation.[2]

---------------------------

The story begins with the supercontinent Pangaea, so named by Alfred Wegener (1880-1930) who promoted a radical idea that he called "continental drift" but is now more accurately referred to as plate tectonics. Pangea was created when the supercontinent Gondwanaland, which consisted of what is now Africa, South America, Antarctica, India and Australia 600 million years ago, collided with what is now Eurasia and North America 480 million years ago. Importantly, the way Pangaea was assembled had what is now southeastern North America connected directly to what is now northwestern Africa (Fig. 6.1).

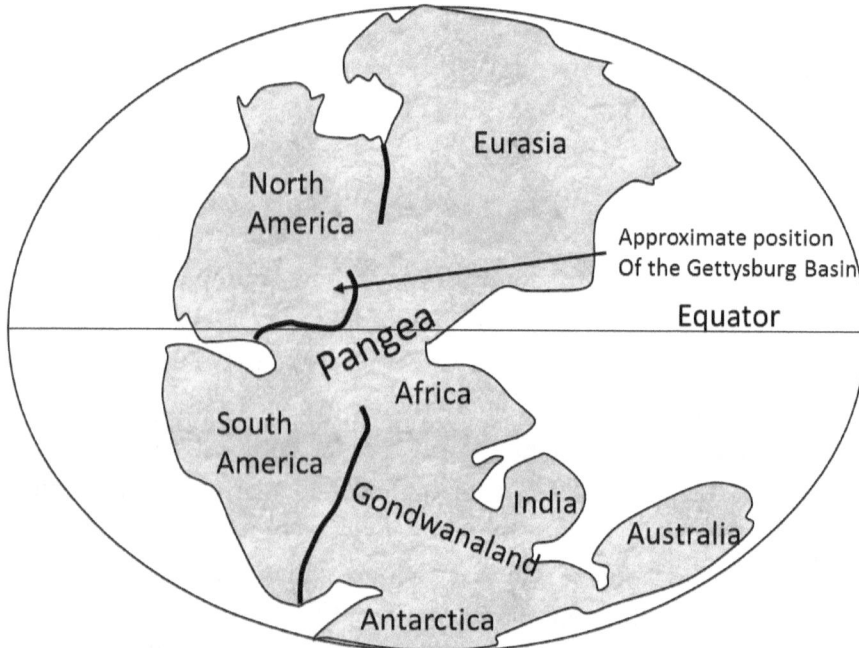

Figure 6.1—The Supercontinent Pangea fully assembled about 250 million years ago.

But beginning 225 million years ago, something else began to happen. The suture welding North America to Africa began to stretch apart. This stretching created faults that caused a series of basins to form roughly parallel to what is now the east coast of

North America. As they subsided, these basins began to collect sediments from the surrounding highlands, and when the climate was sufficiently wet, lakes began to form as well[3]. This pattern of sedimentation was similar to what has been observed in other fault-generated basins like the modern-day Dead Sea. At the lowest part of the basins a coarse conglomerate accumulated rapidly, which you'd expect so near to an active fault zone. However, for the next 20 or 30 million years, the best way to describe the sedimentation is that it was quiet and uneventful. The faulting continued to deepen the basins periodically, but at a gradual pace that was just enough to keep up with the influx of sediments.

One of these basins—now known as the Gettysburg Basin—eventually accumulated sediments to a thickness of about 5,000 meters. Most of the sediments were reddish silts that gradually spread out on broad mudflats in what must have been an arid or semi-arid climate (Fig. 6.2). Also, since this was the beginning of the age of Dinosaurs in what is known as the Triassic Period, the red mudstones that the sediments turned into are famous for having thousands of sets of Dinosaur footprints[4]. Periodically, movement on the faults would deliver coarser-grained red sands that interbedded with the silts, and whenever lakes developed during periods of a moist climate, dark organic-rich clays were deposited as well.

Figure 6.2. Triassic red siltstones and sandstones as seen in a railroad cut at Gettysburg. The pocketknife in the center of the picture shows the scale.

The relatively quiet spell of gradually deepening basins filling with mudflat and lake sediments along the suture between

North America and Africa abruptly came to an end 200 million years ago at the beginning of the Jurassic Period. The rifting between the North America and Africa plates increased in both intensity and violence. As the crust of the Earth was ripped apart, basaltic lava from deep in the mantle welled up and intruded into the siltstones, sandstones. In some places, the basaltic lava flowed out onto land surface. In other places, particularly in the Gettysburg Basin, the lava cooled while it was still underground, forming a black rock known as diabase. As rifting continued, Africa separated entirely from North America, seawater spilled in to fill the deepening basin, and the Atlantic Ocean was born. The rifting between the North American and African plates, which caused the emplacement of diabase dikes and sills into the Triassic Basins, is still going on today in the middle of the Atlantic Ocean.

The distinctive topography of the land around Gettysburg, with its broad flat farmland interspersed with black rocky ridges and hills, is the end result of this long and involved geologic history. Today, as was the case when the Battle of Gettysburg was fought, the flat farmlands are underlain by relatively soft, easily weathered siltstones or sandstones. In contrast, the rocky ridges are underlain by the harder more resistant diabase sills and dikes. The irony of this geologic history has not been lost on historians.[2] The Gettysburg Basin was formed because of the violent separation of North America from Africa. And, of course, the Battle of Gettysburg was fought because of the violent separation of the Confederacy from the Union.

But more importantly, most of what happened during the Battle of Gettysburg was dictated by the geology of the area. On the first day, the Union Cavalry commander immediately recognized that the high ground of Seminary Ridge, which was formed by a diabase dike, would be a strong defensive position and occupied it accordingly. That is the primary reason the coming battle between the Army of Northern Virginia and the Army of the Potomac happened at Gettysburg in the first place. On the second day, Lee ordered General Longstreet's Corp to attack the left flank of the Union line. While marching southward, Longstreet recognized with dismay that if he proceeded on the east side of Seminary Ridge, his advance would be observed by the Union soldiers who occupied Cemetery Ridge, which was a diabase sill (an intrusion of basaltic lava into the sedimentary rocks that is thicker than a dike). Not wanting his movements observed, Longstreet ordered a countermarch to the other side of Seminary ridge so his advance wouldn't be observed. This delayed his assault until the afternoon which allowed the Union troops to position themselves in the relatively flat wheat field and peach orchard, underlain by the siltstones of the Gettysburg Formation. That fateful delay, dictated by the relative positions of the Seminary Ridge dike and the Cemetery Ridge sill (Fig. 6.3), contributed to the failure of the Rebel assault. Also, because the diabase made it impossible to dig trenches or foxholes, it also contributed to the very high casualties suffered by both sides.

Figure 6.3—The diabase on the top of Cemetery Ridge marking "the high water mark of the Confederacy" where Longstreet's Assault was halted.

Which brings us to Longstreet's Assault on the third day. Following a fierce cannonade which the Confederates aimed at the center of the Union line, Longstreet formed up his infantry on the flat ground at the foot of Seminary Ridge (the siltstones of the Gettysburg Formation), aimed them at a clump of trees at the crest of Cemetery Ridge (the diabase sill), and gave the fateful order to attack. This gave every Union cannon on Cemetery Ridge a clear shot at the Confederates, and they inflicted horrifying casualties on the advancing infantry. Amazingly, some of the Confederates managed to reach the diabase rocks underlying the clump of trees (Fig. 4.3), a feat of human courage that is almost incomprehensible, and for a few minutes the Confederates had victory tantalizingly within their grasp. A swift counterattack by Union infantry, however, halted their advance and the battle was effectively over.

-----------------------------------

The young girl's father made a point of getting the family out of the station wagon and walking to the area where the troops of Longstreet's Assault had formed their ranks. Several other visitors were there as well, and everybody gazed uphill in the direction that the assault had gone. One of the visitors said, to no one in particular, that he couldn't believe that the Confederates had dressed ranks right out in the open where the Yankees could see them. Another visitor, apparently a military man, replied that having the men form ranks was the only way that you could get soldiers to move under conditions of such extreme stress. With that, the father shooed his family back to the station wagon and they began the long, hot trip back to Bethesda.

In later years the young girl would remember the miles and miles of cornfields that surrounded the battlefield, and the odd-looking rocky hills that, in places, rose up around the fields. But, like most of the thousands of people who visit Gettysburg every year, she was completely unaware of what those rocks were, why they were there, and how it affected course and outcome of the desperate battle.

For the young girl, that awareness would come later.

# REFERENCES

1. Cuffey, R.J., Inners, J.D., Fleeger, G.M., Smith, R.C., Neubaum, J.C., Keen, R.C., Butts, L., Delano, H.L., Neubaum, V.A. and Howe, R.H., 2006. Geology of the Gettysburg battlefield: How Mesozoic events and processes impacted American history. Field Guides, 8, pp.1-16.

2. Smith, R.C. II and R.C. Keen. 2008. Regional rifts and the battle of Gettysburg, *in* Fleeger, G.M. (ed), Geology of the Gettyburg Mesozoic Basin and Military Geology of the Gettysburg Campaign, 72rd Annual Field Conference of Pennsylvania Geologists, Gettysburg, PA, p. 48-53.

3. Lindholm, R.C., 1978. Triassic-Jurassic faulting in eastern North America—A model based on pre-Triassic structures. Geology, 6(6), pp.365-368.

4. Santucci, V.L. and Hunt, A.P., 1995. Late Triassic dinosaur tracks reinterpreted at Gettysburg National Military Park. Science, 15(1):.9.

# CHAPTER 7.
## A LANDSCAPE IN PARADISE

When he was in the 9th Grade, the young boy took a class that was an introduction to science. The class consisted of several different modules designed to introduce the basics of physics, chemistry, and biology. The young boy liked the class primarily because he liked the teacher, a short stout woman in her 40s named Mrs. Harder. Mrs. Harder was one of those rare teachers who could combine strict classroom discipline with an ability to bring the subject matter to life. For example, when she was introducing the concept of chemical reactions, she set a beaker full of water on her desk at the front of the room, took a fingertip-sized chunk of metallic sodium from a jar using tongs, and dropped it into the water. It bubbled for a few seconds, and then it exploded in a flash of flame, spewing a geyser of water out of the open beaker. *That,* she informed the stunned but now fully attentive class, *was a chemical reaction.*

So it was a great class, and the young boy, who as a rule was an indifferent and habitually inattentive student, found himself intrigued and actually paying attention. One of the modules of the class was entitled Earth Science and involved looking at and identifying samples of rocks and minerals. He learned the difference between igneous, metamorphic, and sedimentary rocks and, remembering West Point, realized for the first time that the steep rocky ridge behind his house was actually made of granite, an igneous rock.

But something else that the class covered also got his attention. According to the book they were using, you could classify landscapes based on their ages—young, mature, or old age. That made immediate sense since, after all, people were also classified as to whether they were young, mature, or old. Even more, a "young" landscape developed after some process or processes had combined to lift the land surface upwards, and the rivers and streams were fast and vigorous and full of energy, just like young people. After the landscape had eroded for a long period of time, however, the slopes of the rivers and streams became more moderate and carried the water slower with much less energy, just like mature people. Finally, after an even longer period of time, the whole landscape eroded down to a flat plain, and the streams and rivers shuffled along slowly barely moving, just like old people. That, thought the young boy, made sense.

A few weeks later, while the boy was at home one Saturday afternoon, a strong thunderstorm blew up and the rain outside poured out the sky. Watching the storm though a window, he suddenly noticed that the sheets of water were collecting into a shallow swale that ran through his backyard, and the resulting stream was rapidly eroding the soil and the beginnings of a gully began to appear. *Cool* the boy thought *it's a young, rapidly eroding landscape!* Not bothering to put on a raincoat, the boy ran outside to watch closer. Sure enough, he could actually see the new gully steadily become deeper, *just like the book had said.* But then something went wrong. As abruptly as the rain had begun, it began to slow down and stop. But the boy wanted to see more, so as the rain slowed down, he fetched a garden hose, turned on the water and directed it down the ever-deepening gully. And it worked. The erosion continued to enlarge and deepen the gully until it was at least a foot deep. Although he was thoroughly soaked by the rain, the young boy nevertheless felt a warm satisfaction. So little of what he heard in school seemed to apply to the real world. But here he had seen a young landscape evolving right before his eyes.

Warm satisfaction, however, was not the response the young boy's father had when he got home and saw the gully. It must have been quite a storm, he said sourly, to have carved such a deep gully so quickly. Filling it in and repairing the damaged grass would be a lot of work that the father was not looking forward too. Hearing his dad's irritation, the young boy judiciously never mentioned the role that the garden hose had played in the "evolution" of the backyard's landscape.

-----------------------

Grand Unifying Theories (GUTs) have always been important in science. The ones we use today—Newton's laws of motion (physics), atomic theory (chemistry), evolution (biology)— are remembered because they were and are useful for making sense of the world around us. But there is another, and surprisingly common category of GUTs that have made, shall we say, less glorious contributions to science. That a GUT is widely accepted, scientifically prestigious, and enormously popular in its time is no guarantee that its veracity will stand the test of time. In some cases, sadly enough, a GUT can have the effect of leading the science straight to a dead end. In geology, there is no better example of this than the GUT, widely popular in the 19th and early 20th centuries and the one that so enamored our young boy, that described how landscapes "evolved" from being youthful to mature and finally to old age.

Historically, it was the exploration of the world's largest and deepest "gully"—the Grand Canyon—that started geologists in America pondering about how landscapes evolved and changed over time. In 1869, John Wesley Powell, who had lost his right arm at the Battle of Shiloh in 1862, led an expedition to describe and map and Colorado River. Powell had grown up on the Mississippi River and was an experienced waterman. He was also a trained geologist, having been a professor of geology at Illinois State Normal School prior to the Colorado River expedition. Ten men, all of whom were experienced in wilderness travel and in the handling of boats, began the expedition. However, only six completed the journey, the other four having left because of the extreme danger and hardships they encountered. In fact, when Powell completed the journey, he was startled to hear that newspaper accounts of the expedition reported that only one man (not Powell) had survived the trip, and that his obituary had been widely published. In the preface to his report of the trip[1], Powell wrote:

> *A good friend of mine had gathered a great number*
> *of obituary notices, and it was interesting and*
> *rather flattering to me to discover the high esteem*
> *in which I had been held by the people of the United*
> *States. In my supposed death I had attained to a*
> *glory which I fear my continued life has not fully*
> *vindicated.*

Apparently, in addition to being an excellent geologist, Powell also had a sense of humor.

As a geologist, Powell could not help but be impressed by the sheer magnitude of the erosive forces generated by the Colorado River flowing though the Grand Canyon (Figure 7.1). Two things were immediately obvious to Powell. First, the Grand Canyon could only have formed if the land including and surrounding the Colorado River had somehow been lifted upward more than a mile sometime in the recent past. Secondly, this uplift, and the resulting increase in the slope of the river, unleashed the erosive forces that carved, and continue to carve, the river's gorges. As Powell considered just how much rock and sediment had to be removed to form the canyon, he couldn't help but imagine what the landscape would look like millions of years from now. Clearly the downward erosion would continue. But also, the steep slopes of the canyon's walls would also erode laterally. Taking that thought to its logical conclusion, Powell could imagine the landscape eroding down to what he termed a *base level*, where there was little or no vertical relief and erosion would become less and less evident.

Figure 7.1—An aerial view of the Grand Canyon. Photo by LT. Kirk McKay, U.S. Navy.

In addition to being a good geologist, Powell also was a fine administrator (a very rare combination of skills). So, when he initiated a geologic survey of the Rocky Mountain region in 1874, Powell recruited a geologist named Karl Grove Gilbert to help with the work. Gilbert, like Powell, had been born and raised in the forested eastern United States, and was equally awestruck by the wide-open landscapes of the West. And also like Powell, Gilbert was forced to consider the forces and processes that sculpted those landscapes. As he worked, mapping and describing the Henry Mountains of Utah, Gilbert expanded on Powell's concept of the base level, and combined it with the thermodynamic concept of equilibrium. What processes, Gilbert wondered, would affect the slope of a riverbed? It was no secret that a steeply sloping riverbed caused more erosion than a shallow sloping riverbed. Why? Gilbert reasoned that the grade of a river must reflect the balance between the energy of the flowing water, and the energy needed to carry the sediment load. When the available energy matched the sediment load being carried, a dynamic equilibrium had been reached and the slope of the river bed could be said to be *graded*. So, in addition to qualitative notion of erosion wearing down a landscape over geologic time, Gilbert's contribution was the realization that the processes involved could be quantified.

By the 1880s, the work of Powell, Gilbert, and others had demonstrated how uplift and erosion could sculpt a landscape, the Grand Canyon being the prime example. But it was a geography professor at Harvard College named William Morris Davis who had what he thought was an astonishing insight. Darwin's theory of evolution had been published in 1859 and it ignited a storm of controversy. Part of that storm, and one that Darwin himself detested, was for various crackpots to use the concept of natural selection to justify the practice of laissez-faire capitalism, racism, warfare, colonialism and imperialism. William Davis was no crackpot, but he became enamored of the idea, based on the observations of Powell and Gilbert, that *landscapes, just like species, evolved over time.* This theory of landscape evolution, which Davis called the *geographical cycle* or the *cycle of erosion*, was first published in 1889 and it immediately became widely popular in America[2].

Davis' cycle of erosion begins with the rapid uplift of a land mass, just as Powell had documented in the land surrounding the Grand Canyon. This rapid uplift was followed by a long period of quiescence during which the landscape evolved through a sequence of stages he termed youth, maturity, and old age. The youthful stage is characterized by high uplands carved with narrow steep valleys. Over time, the landscape moved into maturity. The valleys broadened as erosion removed sediment, and the river slopes reached a graded condition, just as Gilbert had hypothesized. As the landscape continued toward old age, the valley walls gradually flattened out completely, leaving a nearly level plain that Davis called a *peneplain*. The concept of the peneplain was essentially a renaming of Powell's concept of a *base level*. Then uplift would occur again and the cycle would repeat itself. This was the theory of landscape evolution that so enthralled the young boy in his 9<sup>th</sup> grade introduction to science class, and which was responsible for his carving a gully in his father's backyard with a garden hose. It was indeed a Grand Unifying Theory.

It's interesting to speculate about the reasons for the popularity of Davis' theory. One reason for sure is that the idea of "evolution", which many people at the time equated with "change", was simply the hot new concept of the day. In addition, describing landscape *evolution* appealed because it raised the study of landscape morphology, which by the 1890s began to be called *geomorphology*, to the same level of prominence as biological science. Another reason was that it anthropomorphized the study of landscapes into something very familiar to humans: youth, maturity, and old age. Those were concepts that easy for people to identify with and understand.

But another reason had to do with Davis himself. Prior to publishing the cycle of erosion, Davis had been an obscure instructor at Harvard College, at the very bottom of the academic pyramid. After publication, he quickly became the leading figure in geography (Davis was a geographer, not a trained geologist) in the United States. Part of this was due to the fact that the cycle of erosion truly was a marvel of synthesis, neatly tying together geology, biology, and geography into a plausible and appealing framework. But also, Davis had a keen understanding of what it took to rise to prominence in academia. Davis was well-spoken and performed well in scientific meetings. He carried himself with much dignity and was a founding member of the Association of American Geographers in 1904. But Davis also refused to accept any criticism of his theories, often reacting scornfully and derisively to anyone who expressed skepticism. Quite simply, he cowed any opposition to his ideas, at least in America. By the time of his retirement from Harvard, Davis' theory of the cycle of erosion dominated American geomorphology.

Suffice it to say that time has not been kind to Davis' grand unifying theory. After his death in 1934, and thus without Davis there to defend it, it gradually lost steam. There were several reasons for this. First of all, although the cycle of erosion was vividly descriptive, *it was never predictive.* The hallmark of a good hypothesis is that it makes predictions that can be tested by observation or experiment. Davis himself seems to have realized this when he commented[3]:

> ...*the scheme of the cycle is not meant to include any actual examples at all, because it is by intention a scheme of the imagination and not a matter for observation.....*

If it's not a matter for observation, then what can it possibly be used for?

But more importantly the cycle of erosion theory, because of its attractive completeness, actually hid the fact that geomorphologists were woefully ignorant of the physical processes that caused landscapes to change in the first place. Davis' theory is not the only one that suffered from this shortcoming. Consider, for example, Sigmund Freud's theory of psychoanalysis. It too was an attractively complete description of how the unconscious mind governs much of human behavior. And like Davis' theory, it helped to obscure our ignorance about the physical and chemical processes that actually make the human brain work.

Gilbert's contention that landscape processes were amenable to quantitative description had largely been ignored in the early 20th century. Beginning in the 1940s, however, more attention began to be paid to the processes that affected landscapes. One interesting reason for this was the desert campaigns in North Africa during World War II. In the 1930s, a British Engineer named Ralph Bagnold began using motor vehicles to explore the previously unreachable deserts of Libya and Egypt. These deserts are virtually devoid of any vegetation and are covered by huge wind-blown sand dunes known as *barchans*. First, Bagnold developed a technique for driving in the desert so as not to repeatedly get stuck in the sand. But he also developed an equation, known as the Bagnold formula, describing the transport of wind-blown sand in the desert, and how that built and moved barchans.

The Bagnold formula is an early example of how geomorphic processes could be quantified and used to explain land forms (barchans). But, more importantly to the British in WW II, Bagnold's work greatly facilitated the British army's ability to travel and maneuver effectively in the desert. That, in turn, contributed to the defeat of the German Afrika Corps, commanded by Erwin Rommel, in North Africa. That lesson was not lost on geomorphologists: Understanding geomorphic processes on a quantitative level had real practical value. Modern geomorphology, therefore, has shifted from its beginnings as the broad, overreaching Grand Unifying Theory envisioned by William Morris Davis, to a process-oriented, quantitative science[4].

The Grand Unifying Theory of landscape evolution, as espoused by W.M. Davis, had enormous appeal in its time. A good bit of that appeal can be attributed to its coopting elements of Darwin's theory of evolution and applying them to how landscapes change over time. But certainly comparing landscape evolution to the human stages of life—youth, middle age, and old age, concepts imminently familiar to people—was also a factor. Grand Unifying Theories in general have great appeal because they purport to systematize and explain different observations that, taken just by themselves don't seem to be related. But that appeal can have a downside, as was certainly the case with Davis' landscape evolution theory, which is to create the illusion that we know a lot more than we really do.

----------------------

While the young boy's "experiment" trying to reproduce a "young" landscape in his father's back yard turned out to be a failure, Mrs. Harder's introduction to science class made a deep impression on him. In addition to thinking about how landscapes changed over time, Mrs. Harder insisted on the class memorizing the star constellations in the Northern Hemisphere, the Greek alphabet, and the Mohs scale for the relative hardness of minerals (talc = 1, diamond = 10), things that he would remember for the rest of his life. But more importantly, it had begun to dawn on him that the world might not be the chaotic, disordered, unknowable place that it had seemed to be when he was younger.

Maybe, just maybe, the world could make sense.

## REFERENCES

1. Powell, J.W. 1875. The Exploration of the Colorado River and its Canyons. Dover Publications, Inc., New York, 397 p.

2. Davis, W.M., 1889. The rivers and valleys of Pennsylvania. National Geographic Magazine 1(3):15-73.

3. Davis, W.M., 1905. Complications of the geographical cycle. Report of the 8th Geographical Congress 1904: 150-163.

4. Ritter, D.F., Kochel, R.C. and Miller, J.R., 1995. Process geomorphology, Fifth Edition, Waveland Press, Inc.

# CHAPTER 8.
## MAGIC IN THE NUMBERS

When the young girl began the 11th grade, she was dreading having to take Algebra II/ Trigonometry. She had taken geometry in the 9th grade and Algebra I in the 10th, had thoroughly disliked both, and hadn't gotten very good grades. Part of the problem was that the teachers in her Catholic school—they were all nuns—were not particularly good at explaining math. But the other problem, and this is a trap that many high school kids fall into, was that because math seems to be difficult it becomes distasteful to work problems. That, in turn, leads kids to do as few problems as they can get away with. And because they don't do many problems, it just makes math that much harder to learn. Bad attitudes and bad grades are the usual outcome. So for a lot of kids, disliking math becomes a negative feedback loop, a self-fulfilling prophesy. Our young girl was one of these unfortunates.

So when the class started, she immediately began to have trouble. She'd listen to the teacher lecturing and doing problems on the blackboard, but it just didn't seem to make very much sense. This, she thought bleakly, was going to be awful, she'd probably get an awful grade, and that would make this an awful year.

But then she got lucky. While she was trying to make sense out of what the teacher was saying, she noticed that the girl sitting next to her, whose name was Maria, wasn't listening to the lecture at all. Rather, she had her book open and was simply doing the problems the teacher was laboring to explain. When the young girl asked Maria what she was doing, Maria told her "don't listen to the teacher, she'll just confuse you. Instead, do the problems in the section the teacher is covering. If you get stuck, read the book. It's a lot easier to understand the book." As it happens, Maria's parents were both math professors at the University of Maryland, and that was the advice they had given her years ago. "Open your book and I'll show you how", Maria said.

So the young girl began to imitate the way Maria was working through the problems. At the end of each section in the book there was a numbered list of practice problems. The teacher would work through the even-numbered problems during the lecture. Then, she would tell the students to do the odd-numbered problems on their own. Any problems they couldn't finish in class would become homework. So rather than listen to the lecture, the two girls would work the problems on their own, referring to the book if they got stuck. And it worked. All of the sudden, the formidable logic that underpinned mathematics began to dawn on the young girl.

It turned out to be a life-changing revelation.

Maria and the young girl fell into a routine. First, they would work through the even-numbered problems as the lecture was going on. Then, without waiting for the teacher to finish, they'd start on the odd-numbered problems. Often, they would be done with all of the problems by the time the teacher stopped lecturing. That, however, created a different problem. Because they had finished their homework, the girls did what young girls do. They sat at their desks and chatted. The teacher, noticing this, assumed the girls were goofing off and told them to get to work. "But we're done", Maria said. Not believing her, the teacher called Maria to the blackboard and told her to work problem number 7. Maria promptly did. Frowning, the teacher picked another problem. Maria got that one too.

This became a game of sorts. The girls would surreptitiously do the problems on their own while the teacher was droning away, finish, and then start chatting amiably. Their only concession was to whisper, or if that failed, they'd pass notes back and forth. The teacher never figured out what they were doing, and periodically would call Maria or the young girl to the blackboard to do problems. But Maria's method of studying— doing lots and lots of problems—worked. Instead of Algebra II/Trig being her hardest class, it became her easiest. For the first time in her life, the young girl got an "A" in a math class. But even better, her fear and dislike of math was gone forever.

That, in turn, would change her life.

--------------------

One of the difficulties of studying the Earth and how it works is that so much of what goes on is hidden away from view. We live at land surface, of course, and observing what goes on even a foot or two below land surface is either impossible or highly inconvenient. Perhaps the best example of this is water. We can easily see and observe the water in lakes and streams, and because of this many people assume that that's where most of the fresh water in the world resides. But actually, about 98% of the fresh, liquid water on earth is not found in lakes or steams, but is hidden away as groundwater. Its hidden nature is largely why, for most of human history, groundwater was considered to be so mysterious and magical[1]. But mystery and magic are not the ideal ways to consider something as important as groundwater.

For that, you need math.

Mystery and magic were the last things on the mind of Henry Philibert Gaspard Darcy[2] (1803-1858) in October of 1855. Rather, he had gone to considerable trouble to build an experimental apparatus for measuring how fast water would flow through sand. His motivation for wanting to know this had little to do with scientific inquiry and much to do with practical engineering. Beginning at least in the 6th century AD, it was well-known that filtering water through sand could purify it of sediment and noxious dissolved organic matter. The filter cisterns of Venice and Constantinople, many of which were built during the Middle Ages and are still in operation today, are clear evidence of this understanding. But during the Middle Ages, that understanding was entirely qualitative. The artisans who built the cisterns learned by trial-and-error the correct dimensions and the correct amount and kind of sand to use. But qualitative wasn't going to work for Henry Darcy. As an engineer, he knew from both training and experience that to build an efficient sand filtering plant, he needed to quantify the relationship between the properties of the sand, the dimensions of the filter, and the water pressure. The apparatus he had built was designed explicitly for that purpose.

Water and water quality had been important to Darcy as long as he could remember. When he was a boy growing up in Dijon, France, water was supplied from shallow wells or from the L'Ouche River. Unfortunately, by virtue of the lack of efficient disposal of human and animal waste, much of this water was heavily polluted. Thanks to his mother, Darcy was able to attend the *École Polytechnique* in Paris, the premier engineering college in France. He later transferred to the French School of Bridges and Roads. After graduation, he was offered a job with the Corps of Bridges and Roads, and one of his first projects was to design and build a water supply for his hometown of Dijon. Years later, Darcy would recall just how bad Dijon's water really was[2]:

> *Dijon was in a deplorable situation with respect to potable water. The only water the inhabitants had was from private wells, and about a hundred wells located on the public streets. These public wells were not even covered, and it was not rare that the buckets that brought up the water for domestic use would contain a dog or cat that had been drowned several days previously.*

The city of Dijon had attempted to drill deeper wells in order to get a better water supply, but these efforts failed. Darcy's solution to the problem was to take advantage of the fact that Dijon is located in a river valley at a relatively low elevation. The hills surrounding Dijon, however, not only were at a higher elevation but they had several springs that discharged significant volumes of water. Darcy's system collected water from the Rosoir Spring located eight miles from Dijon, carried it to a reservoir through a covered aqueduct, and then distributed the water throughout the city by a network of pipes. The entire system was gravity driven and so was very economical. But the system also used sand filtration to insure that the water quality was acceptable. While he was building the filtration system Darcy, like the builders of the filter cisterns in Venice, simply used a trial-and-error approach. But as a good engineer, hit-or-miss was not how he liked to approach problems. Because it was impractical to directly *observe* how water flowed through sand, Darcy did the next best thing. He turned to mathematics.

Darcy approached the problem in a very straightforward and practical way. Darcy and his colleague Charles Ritter filled a cylindrical column with sand and equipped it with mercury manometers (pressure gauges) to measure pressure at the top and bottom of the column (Fig. 8.1). In the experiments conducted in October, 1855, they systematically varied the incoming water pressure (by opening or closing a valve connected to a spigot) and measured the discharge of water at the bottom of the column. In addition, Darcy also measured how the discharge varied with the length of the column if the pressure was kept (approximately) constant.

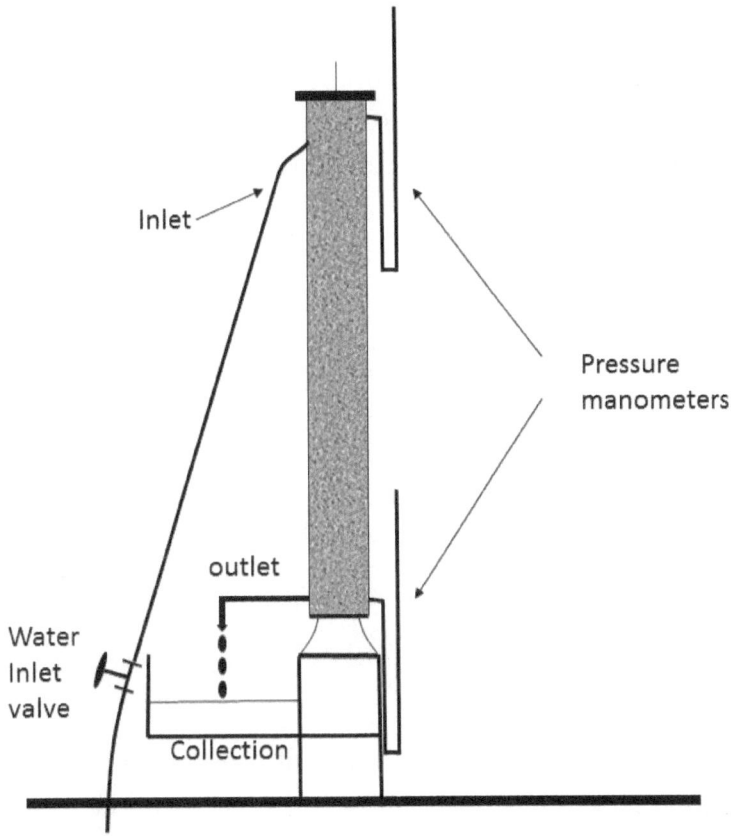

Figure 8.1—Darcy and Ritter's apparatus for measuring water flow through a sand filter[2].

Darcy and Ritter observed two things that they immediately recognized as being significant. First, the discharge of water from the column *increased* in proportion to the increase in pressure from the top of the column to the bottom. Secondly, they observed that if they kept the pressure constant, the discharge decreased in proportion to the length of the sand column. Importantly, Darcy and Ritter also recognized that the force driving the water flow included the pressure added by the spigot (which they denoted as *P*), and the height of the water column (which they denoted as *h*). In their own words[2].

> *Thus, if we denote the thickness of the sand layer by*
> *e, its surface area by s, atmospheric pressure by P,*

*and the height of the water on the sand layer by h,
we will have P + h for the pressure to which the
upper end will be subjected. In addition, if P ± h₀ is
the pressure to which the lower surface is subjected,
k is a coefficient that depends on the permeability of
the (sand) layer, and q is the volume discharged, we
have:*

$$q = k\frac{s}{e}[h + e + h_0] \text{ which reduces to } q =$$

$$k\frac{s}{e}[h + e] \text{ when } h_0 = 0 \text{ or when the pressure under}$$

*the filter is equal to atmospheric pressure.*

They then rearranged the equation by dividing both sides by the surface area *s* to give:

$$\frac{q}{s} = v = \frac{k}{e}[h + e] \qquad \text{(Equation 8.1)}$$

where *v* is the water discharge per unit surface area of the filter, also known as the *specific discharge*. This entirely empirical equation, now known as Darcy's Law, enabled Darcy and Ritter to design filter systems for purifying water without having to resort to trial-and-error.

The modern form of Darcy's Law reflects a couple of changes, one that reflects convenience and another that is more fundamental. If, unlike the Darcy/Ritter experiment, the column is laid flat (i.e. h = 0), the term ($p_2 - p_1$) is always negative for water flowing from left to right in the column (Fig. 8.2). For that reason, it's useful for *k* to be denoted as a negative number. That, in turn, means that the specific discharge is (conveniently) a positive number.

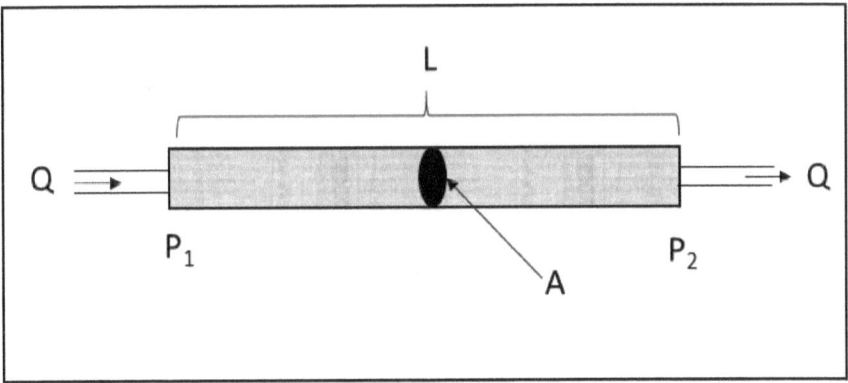

Figure 8.2. Diagram showing the variables of Darcy's Law used in Equation 8.2.

But also, because Darcy and Ritter were concerned only about fresh water, it didn't occur to them that fluids flowing through sand were affected not only by the permeability of the sand, but also by the viscosity of the fluid itself. But importantly, Darcy's Law applies not only to water but to other useful fluids, oil being one example. Furthermore, the viscosity of oil can vary a lot between different oil reservoirs. Clearly, the more viscous the fluid, the more pressure it takes to move it though sand. That in turn means that the specific discharge will be inversely proportional to the viscosity. The modern formulation of Darcy's law, therefore, includes consideration of viscosity. In addition, rather than using $e$ to describe the length of the column, the symbol $L$ is now generally used. This gives Darcy's Law as:

$$\frac{Q}{A} = v = -\frac{k}{\mu L}[p_2 - p_1] \qquad \text{Equation 8.2}$$

where Q is discharge, A is the cross-sectional area of the column, v is the specific discharge, k is the intrinsic permeability of the sand, μ is the fluid viscosity, L is the length of the column, and $p_1$ and $p_2$ are the pressures at the beginning and end of the column. The beauty of Darcy's Law is that it helps us address the behavior of fluids where it is impossible to directly observe them. That, in turn, can prove to be very useful.

------------------------

Paul Hsieh stared at the picture on his cell phone. It showed a pressure gauge attached to the Deepwater Horizon oil well that had suffered an explosion on April 20, 2010, killing eleven people and injuring seventeen. This was 86 days later, and engineers had just managed to cap the well, staunching the flow of crude oil into the Gulf of Mexico. The problem was that six hours after shutting the well, the indicated pressure on the gauge hadn't risen as high as the engineers thought it should. That could mean that the explosion had damaged the well casing causing oil to leak into the surrounding sediments and rocks. That, in turn, could cause the formation to fracture, a fracture that could grow and breach the seafloor. If that happened, the leak might never be stopped. Because of this, some of the engineers were advocating reopening the well to prevent the dreaded formation fracturing from happening.

Paul Hsieh, a Research Hydrologist working for the U.S. Geological Survey, was an expert in modeling fluid flow in groundwater systems. Looking at pressure gage, he wasn't convinced that the pressure in the well was actually too low. Paul was a member of a team of government experts advising the British Petroleum engineers, but at the time he was home in Menlo Park, California. A friend of his who was at the Deepwater Horizon site had taken a picture of the pressure gauge reading and texted it to Paul. It was about six in evening when Paul got the text.

It had previously been determined that if the pressure in the well after being capped rose to 7,500 pounds per square inch (psi), then the risk that there had been major damage to the well was low.

On the other hand, if the pressure was below 6,000 psi, major well damage was likely. If the pressure was between 6,000 and 7,500 psi, however, the condition of the well was simply uncertain. The pressure indicated by the gauge in the picture on Paul's phone read 6,600 psi—dangerously close to the "major well damage" zone.

But Paul also knew that, because the well had been flowing continuously for 86 days, the pressure in the oil reservoir 18,000 feet below the surface of the Gulf of Mexico would have been severely depleted. Was it possible that the low pressure reflected depletion of the pressure due to the loss of oil from the reservoir and not a leak? If so, how would we know?

Paul's answer was to construct a mathematical model of the oil reservoir that was based, in part, on Darcy's law (Equation 8.2). This involved putting in the known dimensions of the reservoir (analogous to $L$, the dimensions of Darcy's column (Fig. 8.3)), the reservoir pressures observed before and after the explosion (analogous to $p_1$ and $p_2$), the intrinsic permeability of the reservoir ($k$ in Equation 7.2), and the viscosity of the oil ($\mu$ in Equation 7.2). All of this was much more complicated than just Darcy's law because the simulations were done in two dimensions (Fig. 8.3 considers only one dimension) and because time had to be part of the analysis (time is not considered in Darcy's Law). Nevertheless, Darcy's Law was a fundamental relation in the computer model that Paul used. In his report of his analysis[3], published more than a year after the well was capped, Paul listed the assumptions he used to construct the model commenting[3]:

*These (assumptions) are standard in the analysis of pressure buildup and flow tests in oil wells, and include assumptions that the reservoir is horizontal, the fluid compressibility is small and constant, and that pressure gradients with the reservoir are sufficiently small for Darcy's Law to apply.*

Paul stayed up all night building and tweaking his model, and by morning had enough results to suggest that the lower-than-expected pressures in the reservoir might not indicate a breeched well casing after all. Instead, they might simply reflect the pressure loss due to the loss of oil over the 86 days. In the morning, Paul communicated his findings to the engineers on site via a WebEx. After much discussion, it was decided to continue monitoring the pressure buildup and not resume discharging oil into the Gulf. As new pressure readings became available, they generally matched the values predicted by Paul's model. The well remained shut until August 3, 2010 when mud and cement was injected into the well beginning the "static kill" that would cap the well for good.

Mathematics had pierced the veil of uncertainty.

------------------------

The young girl, thanks to her friend Maria, learned several important lessons during 11<sup>th</sup> grade. First was the fact that her previous difficulties with mathematics was completely unrelated to her innate abilities. Many people think that there is a "knack" associated with being good at mathematics, and that knack is something that you either have or don't have. That might be true for the high-level math of particle physics, but certainly not for high school algebra. Secondly, and most important of all, she learned that the trick of learning math was simply practice. In order to get good at it you just had to spend time doing problems. If a problem gave you trouble, you go to the book, or to the teacher, to get on the right track. Then do the problem over a couple of times. Then do another similar problem. Then another. It wasn't really any different than learning any other skill like playing the piano, throwing pottery on a wheel, or playing tennis. The trick was simply to practice it over and over again until it became second nature.

Losing her fear and distaste math opened a whole new world of opportunity for the young girl. Her mother, fully aware of her earlier difficulties with math, was counseling her to major in home economics when she got to college. But with her new comfort and interest in mathematics, a whole range of subjects that had always interested her—biology, physics, and geology—now became a possibility. Years later she would remember how thrilled she'd been to get an 'A' in Algebra II/Trigonometry, and how glad she was that Maria had showed her how. That experience stood her in good stead as she began taking the calculus and physics classes that are the basis of any science major. Eventually, when she was taking a class in petroleum geology, she became acquainted with the fundamental equation governing the flow of fluids in the subsurface.

Darcy's Law.

## REFERENCES

1. Bord J., Bord C. 1985 Sacred waters: holy wells and water lore in Britain and Ireland. Brandada publishers, London, pp 1–25.
2. Darcy, Henry, 1856. The public fountains of Dijon. English translation by Patricia Bobeck, 2004. Kendall/Hunt Publishing Company, Dubuque, Iowa, 506 pp.
3. Hsieh, P.A., 2011. Application of MODFLOW for oil reservoir simulation during the Deepwater Horizon crisis. Ground Water, 49(3), pp.319-323.

# CHAPTER 9.
## CARVED IN STONE

The summer before the young boy started his senior year of high school, his father retired from the Army and the family moved from Virginia to a town called Cockeysville, Maryland. In 1725, a man named Thomas Cockey had settled in what was described as a "limestone valley" a few miles north of Baltimore. Over the years, the Cockey family farmed the land and engaged in various commercial pursuits that included building a hotel in 1810. That hotel became the nucleus of a little town that the locals named Cockeysville.

You might think that having to change schools for his senior year would have been difficult for the young boy. But, because the Army had been moving his father from post to post for as long as he could remember, he didn't think it was a big deal. Since kindergarten, he had attended eight different schools by the time he finished the 11th grade. He wasn't happy about changing schools one more time, but it wasn't as if this was something new. He was used to it.

One of the perks of being a senior at his new school, however, was that he could drive his old, beat-up Volkswagen bus to school. This involved driving west from Cockeysville on Padonia Road which, just before it crossed underneath I-83, dipped into a shallow valley. That valley, one of the "limestone valleys" of Thomas Cockey, held what looked like several fairly large ponds. When he looked closer, he realized that the "ponds" were actually the excavations of old quarry workings that had subsequently filled up with water. It didn't look like the quarry was still active, in fact it looked like it had been abandoned for some time. He wondered what had been quarried there and how long it had been active. As it happens, this had been a marble quarry since about 1834, and between 1845 to 1884 this quarry produced most of the marble used build Washington, D.C.'s most recognizable landmark.

The Washington Monument.

The Cockeysville Marble, as the stone is known to builders and geologists alike, has been described as[1]:

> ....*a completely crystalline white marble of variable grain. It varies from a fine-grained marble to a coarse rock known as alum stone. Certain beds are gray, pink or pinkish brown in color. The beds vary greatly in thickness, but the average depth of the strata may be said to be 400 feet. The pure white marble is the best and the one most highly desired for building purposes.*

In addition to being used in building the Washington Monument, the Cockeysville Marble has been used for a variety of monuments throughout Maryland. One of these is a marking stone memorializing the more than 3,000 Confederate soldiers who died at a Civil War prison camp located near Point Lookout, Maryland. You can see from that marking stone (Figure 9.1) that it is indeed *a crystalline white marble of variable grain*, and the stone itself is very beautiful.

Figure 9.1—Stone marker for the Confederate Cemetery at Point Lookout, Maryland, made from the Cockeysville Marble.

So it's no particular wonder that the Cockeysville Marble was the building stone of choice for the Washington Monument, and for dozens of other buildings in Washington D.C. and Baltimore. The comment that "the pure white marble....most highly desired for building purposes" might give the impression that the rock is fairly uniform. But, as you can see from the marking stone (Fig. 9.1) the Cockeysville Marble, like almost all quarried building stone, is not uniform at all. The white marble that you see in the Washington Monument is more a reflection of the quarryman's skill in selecting and carefully carving out the most desirable parts of the stone. To the quarryman, variability in the stone was a nuisance that had to be dealt with like any other business problem. To the geologist, however, that variability contains clues as to how the rock was formed in the first place, and why it acquired the desirable properties that it has.

--------------------------------

In 1905, a geologist named W.J. Miller[2] measured and described a section of the Cockeysville Marble, probably in an active quarry. Miller's description reads like this, starting at the top of the section:

## SECTION SHOWING ALTERNATIONS OF CALCITIC AND DOLOMITIC MARBLE

| Rock Description | Thickness | |
|---|---|---|
| | feet | inches |
| Medium grained, calcitic...................... | 5 | |
| Rather coarse grained, clear, white, calcitic.. | 1 | |
| Coarse grained, bluish, pyrite, calcitic........ | | 2 |
| Very fine grained, friable, dolomitic.......... | | 4.5 |
| Very pure, coarse grained, calcitic............ | | 1.5 |
| Fine grained, gray, micaceous dolomitic..... | | 10 |
| Fine grained, grayish brown, impure, calcitic | | 4 |
| Fine grained, pure, dolomitic.................. | | 6 |
| Medium grained, blue, calcitic................. | 1 | 8 |
| Medium to coarse grained, white, calcitic.... | | 10 |

This is just the top ten and a half feet of a 73-foot section, but it's enough to illustrate the point that most of what geologists call the Cockeysville Formation is *not* made up of the pure white marble you see in the Confederate Marker (Fig. 9.1) or the Washington Monument. More typically, the metamorphic rock is layered into discrete beds of sedimentary origin that differ markedly in both appearance and in mineralogy.

There are several things about the above description that are worth pointing out. First, each of the layers are described as either "calcitic" or "dolomitic". This refers to the two minerals that make up most carbonate rocks. The first is the mineral *calcite* that has the chemical formula $CaCO_3$, and the second is the mineral *dolomite* that has the chemical formula $CaMg(CO_3)_2$. Calcite, therefore, consists entirely of calcium ($Ca^{2+}$) and carbonate ($CO_3^=$) ions packed together in a crystalline structure. Dolomite, on the other hand, has equal portions of calcium and magnesium ($Mg^{2+}$) ions packed together with two molecules of carbonate. Because magnesium has a slightly smaller diameter (160 picometers) than calcium (197 picometers), the ions in dolomite arrange themselves into a slightly different structure than calcite. That "slight difference" in crystal structure had an important role in producing the "white" marble.

Another thing to notice is that several of the rock layers are described as being "impure", or "micaceous" meaning that there were other minerals present in the rock other than just calcite and dolomite. Finally, the "calcitic" layers are mostly described as either being "medium grained" or "coarse grained", whereas the "dolomitic" layers are all described as being "fine grained". That suggests that something different happened to the calcitic layers of rock relative to the dolomitic layers. But what?

A clue to that, ironically, lies in the masses of pure, white, calcitic marble beloved of the quarryman and builders alike. These marble "masses" do not occur in discrete beds, as in Miller's log above. Rather, they occur in huge irregular bodies embedded in the surrounding beds of metamorphosed dolomite[3]. Furthermore, the marble masses cut across the sedimentary layers that are still visible in the metadolomite (Fig. 9.2). In order to explain that, we have to go back to how carbonate rocks, limestones and dolomites, form in the first place.  Then we can see how from those carbonate rocks were transformed into marbles.

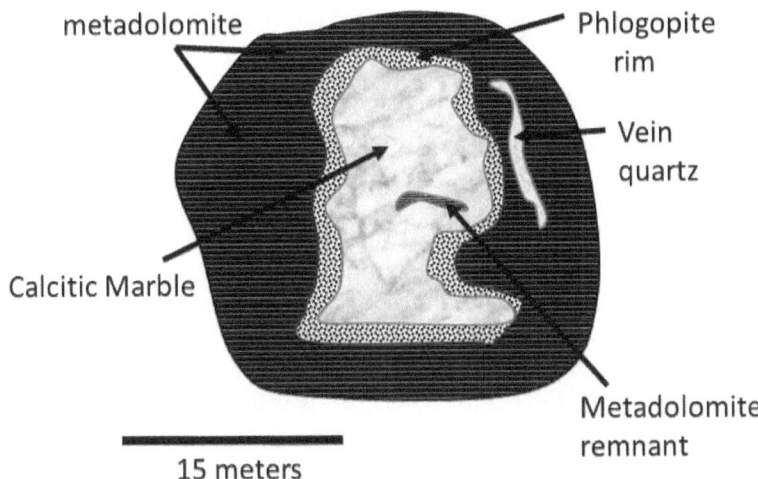

metadolomite

Phlogopite rim

Vein quartz

Calcitic Marble

Metadolomite remnant

15 meters

Figure 9.2—Schematic diagram showing the occurrence of metadolomite, calcitic marble, phlogopite rims, vein quartz, and metadolomite remnants within the Cockeysville Formation. Modified from Choquette, 1960.[3]

Carbonate rocks made up of calcite and/or dolomite typically begin life as soft, spongy sediments in marine or lacustrine environments.  In modern carbonate sediments, the principal source of the carbonate are clams, snails and other shell-making organisms.  This carbonate shell material is subsequently ground up into smaller grains by wave and tidal action before being deposited.  But carbonate minerals, particularly dolomite, can also be produced by a variety of microorganisms as a byproduct of their metabolism.  The Cockeysville Marble, therefore, had its beginnings with the production and deposition of carbonate material, probably over several million years, in a marine environment.  The relative lack of clastic sediments like sand, silt, or clay suggests a marine environment that was not particularly near any clastic sediment-producing land mass. The age of the Cockeysville Marble, which is about 500 million years, suggests it formed in a shallow sea when North America was separated from the supercontinent of Gondwana (see Figure 5.5A).

After a long period of carbonate deposition, which deposited anywhere from 100 to as much as 200 meters of carbonate sediment, things began to change in the tectonic environment. A sudden influx of clastic sediments washed in from volcanic arc islands approaching North America and buried the carbonate sediments. When the volcanic arc islands collided with North America (see Fig. 5.5B) causing the Taconic Orogeny, both the carbonate and clastic sediments were subjected to extreme heat and pressure. This began the process of turning the clastic sediments into schists and gneisses, and turning the alternating layers of calcite and dolomite into marble. These metamorphic processes were repeated when Gondwana later collided with North America, forming he supercontinent Pangea 320 million years ago (Fig. 6.1).

This metamorphism, which went on intermittently over 100 million years, was not uniform. Much of the Cockeysville Formation simply remained a fine-grained layered metadolomite (Fig. 9.2). But in some places, probably starting in the calcitic layers, the carbonate material began to recrystallize. This recrystallization seems to have spread to the dolomite layers, removing the magnesium and turning the rock into a remarkably pure calcite. This recrystallization also seems to have removed many of the clastic impurities that were in the dolomites, probably because they wouldn't fit inside the tight crystal lattice of the calcite. These exported impurities, including silica and clay minerals, migrated out of the calcite and were re-deposited on the outer fringes of the pure white marble masses as veins of quartz and a mica-like clay mineral called phlogopite (Fig. 9.2). A raft of other trace minerals were also formed including diopside, sphene, apatite, capolite, pyrite and limonite. The association of these metamorphic minerals suggests the Cockeysville Formation was subjected to temperatures up to 600 °C and 6 kilobars of pressure[4]. That, in turn, suggests the Cockeysville Formation had been buried to a depth of about 25 kilometers while the marble was actively forming.

The masses of recrystallized pure white calcite in some cases grew quite large, as much as twenty or even thirty meters thick. In some cases you can see remnants of layered dolomites embedded in the calcite masses indicating that the marble was in fact altered from the dolomite. In addition, you can also see how the layering of the surrounding metadolomites extends on either side of the recrystallized calcite marble (Fig. 9.2). These large masses of recrystallized calcite are what quarrymen and builders think of as the Cockeysville Marble, even though they represent a small minority of the total volume of the Cockeysville Formation.

During the 19th and early 20th centuries, many of the larger marble masses were quarried and used as building stone. The big customer in the 19th century, of course, was the Washington Monument. In the 20th century, the marble was shipped all over the eastern United States. In 1929, for example, marble from the Beaver Dam Quarry (which is now just a big pond) was chosen as one of the stones to be used to decorate the Fisher building in Detroit, Michigan[1]. The building, which is still a landmark skyscraper in downtown Detroit, was financed by the Fisher family when they sold their company, Fisher Body, to General Motors in the 1920s.

Other than the Washington Monument, the most visible use of the Cockeysville Marble as a building stone can still be seen in the ethnic neighborhoods of downtown Baltimore. The row houses in Highlandtown (Polish), Little Italy (Italian), and Greektown (Greek) just to name a few, were often have steps made of the Cockeysville Marble. These white steps stand out distinctly against the red brick used to build the rowhouses, and they are still quite beautiful. The owners of the row houses take great pride in the beauty of their marble steps and keep them scrupulously clean, often sweeping them daily.

The geological definition of a "rock" is a naturally occurring aggregate of lithified material composed of more than one mineral. The term "stone", on the other hand, refers to a rock that has properties that make it desirable as a building material, usually beauty and durability. Beauty and durability, in turn, often reflects something rare and unusual in a rock's history. In the case of the Cockeysville Marble, that included millions of years of heating and squeezing a dolomitic sedimentary rock, turning some of it into the coarse-grained, calcitic, tightly-latticed, nearly pure white Cockeysville Marble.

---------------------------------------------

For one year of his life, his senior year in high school, the young boy drove by the water-filled remnants of the Cockeysville Marble quarries going to and coming home from school. It wasn't a very happy year. Most of the people in his high school had been together since kindergarten, and the friendships and cliques among the students had developed and cemented long before he came along. It wasn't horrible, but it also wasn't much fun. And the fact that it wasn't much fun had one positive effect. It made him very much look forward to next year.

When he could go to college.

## REFERENCES

1. Purdum, W.D., 1940. The history of the marble quarries in Baltimore County, Maryland. On file in Maryland Room, Hornbake Library, University of Maryland, College Park.
2. Mathews, E.B. and Miller, W.J., 1905. Cockeysville marble. Geological Society of America Bulletin, 16(1), pp.347-366.
3. Choquette, P.W., 1960. Petrology and structure of the Cockeysville Formation (pre-Silurian) near Baltimore, Maryland. Geological Society of America Bulletin, 71(7), pp.1027-1052.

4. Sanford, R.F., 1980. Textures and mechanisms of metamorphic reactions in the Cockeysville Marble near Texas, Maryland. American Mineralogist, 65, pp.654-669.

# CHAPTER 10.
## CATASTROPHE OR OPPORTUNITY

For the young girl, now a young woman, college came as a bit of a shock. For one thing, from 1$^{st}$ to 12$^{th}$ grade, she had attended a Catholic girl's school. So simply seeing men on a daily basis was a new and distracting experience. For another, there was the inevitable acclimation to college life which in her case involved attending Montgomery College, a community college in nearby Rockville, Maryland. Money being tight in her family, it was judged that starting out in a community college and living at home was the most economical way to begin her college career. Finally, and again because money was tight, it also meant having a part-time job. Getting used to college and working her job in the Men's Work Clothing Department at Sears and Roebuck, took some getting used to.

Then there was the problem of deciding what she should major in. She was interested in math and science, but which science should she pick? In her first semester, she took a class in introductory biology which turned out to be fascinating. The professor was energetic and engaging, and they covered the basics of species, phylogeny, physiology, genetics, evolution and ecology. She loved it, and it looked like biology was going to be her major. So the next semester she signed up for a class in ecology, the next course in the sequence leading to a degree in biology. But, because she had also taken and liked an Earth Science class in high school, she also signed up for an introductory geology class.

For the first several weeks of the semester, it seemed that the ecology and geology classes were separate, distinct, and unrelated. The ecology class focused on the ways that different species interacted with each other, how ecosystems worked, and how nutrients and organic carbon were cycled through those ecosystems. The geology class, on the other hand, focused on rocks and minerals, the structure of the earth's crust, and the new theory of plate tectonics.

But about halfway through the semester, the ecology class shifted focus from how ecosystems *worked* to the *damage* humans were doing to them. Habitat destruction, chemical contaminants, over-hunting and over-fishing were all contributing to ecosystem damage and, inevitably, to the extinction of species. At about the same time, however, the geology class introduced the geological time scale. To her surprise, she realized that while the ecology class viewed species extinction as a calamity, something terrible that dealt irrevocable damage to the earth, the geologic time scale was actually *based* on the periodic extinctions of some species and the appearance of new species. These extinction/appearance events had occurred over and over in the history of the earth, and the fossil record of these events served as a convenient way to separate and mark different intervals of time. The contrast between the ecology class view (species extinction as ecological disaster) and the geology class view (species extinctions as periodic events) could not be starker.

That got her thinking.

------------------------------

Prior to about 1750, the idea that any species could become extinct, or that new species could also appear, was so outlandish that it simply never occurred to anybody. The reason is simple enough. The Genesis story in the Bible is clear that all of the species in the sea, and air were created on the fifth day of creation:

> *And God said, "Let the water teem with living creatures, and let birds fly above the earth across the vault of the sky."* Genesis 1:20

Similarly, the creatures on the land were created on the sixth day:

> *And God said, "Let the land produce living creatures according to their kinds: the livestock, the creatures that move along the ground, and the wild animals, each according to its kind." And it was so.*

105

Genesis: 1:24.

For centuries, the Bible had been the only document available in Europe that specifically addressed how the world was created. If a species became extinct, that would mean that God was destroying His own creations, which didn't make any sense. In the same biblical context, if God had wanted new species, he would have created them on either the fifth or sixth day. So there was no reason or mechanism for new species to appear.

Georges-Louis Leclerc, the Count of Buffon (1707-1788), a French naturalist and cosmologist, made one of the first attempts to construct a more rationalist explanation for the existence of differing species. In the 36 volumes of his treatise *Histoire Naturelle*, Buffon suggested that some species had been "improved" following their creation whereas other species had "degenerated". He made the interesting suggestion that all of the four-legged creatures (many hundreds of species) on earth were derived from just 36 of God's original species. The mere suggestion that species could be "transformed" into other species was an astonishing idea and one that was promptly condemned by the Catholic Church. Buffon dutifully recanted in order to deflect the political heat, but he didn't stop writing about his ideas.

Buffon was also skeptical of the purported date of creation as calculated by Archbishop James Ussher from the family lineage recorded in the Bible. Ussher's calculations placed the exact date of creation on Saturday October 22, 4004 BC at 6 PM in the evening. Buffon, using measured rates of how fast spheres of iron cooled, calculated that the age of the earth had to be at least 75,000 years. Once again the Church objected, once again Buffon recanted, and once again he continued writing his volumes on natural history. Buffon was a talented writer, and his volumes were widely read in Europe. The seeds of what would become the theory of evolution had been planted.

The first person to actually prove that a species could become extinct was the zoologist Georges Cuvier (1769-1832).

Cuvier was studying the fossilized skeletons of elephants that had been found near Paris. By carefully examining the anatomy of the fossilized skeletons he was able to show conclusively that they differed from the bones those of existing African or Indian elephants. Ergo, they must have belonged to a different species. Furthermore, since no living examples of the fossilized elephants existed, and since it would be pretty hard to miss them if they did exist, it followed that they must be extinct.[1] By applying this classic sequence of Enlightenment observation and reasoning, Cuvier showed that the concept of extinction was no longer just speculation. It was a fact.

Cuvier also studied the marine fossils found in the Paris Basin, and he observed that the fossil assemblages often changed abruptly from sedimentary layer to sedimentary layer. The only way he could reconcile that observation was to envision cycles of creation and subsequent destruction of marine organisms by catastrophic events. Cuvier wrote:[1]

> *This change was sudden, instantaneous not*
> *gradual, and that which is so clearly the case in this*
> *last catastrophe in not less true of those which*
> *preceded it.*

Cuvier reasoned that these catastrophic events, such as world-wide deluges, were capable of wiping out whole swaths of organisms:

> *Life upon the earth in those times was often*
> *overtaken by these frightful occurrences. Living*
> *things without number were swept out of existence*
> *by catastrophes.......The evidences of those great*
> *and terrible events are everywhere to be clearly*
> *seen by anyone who knows how to read the record*
> *of the rocks.*

As such, Cuvier became a proponent of *catastrophism,* the idea that catastrophic events were responsible for the sudden changes in fossil assemblages between different stratigraphic layers.

But Cuvier's catastrophism did not fare well against the

*uniformitarianism* espoused by James Hutton and later by Charles Lyell in the 19th century. Hutton, and especially Lyell, preferred to think of geologic processes—deposition, uplift, erosion—as happening slowly and continuously over long periods of time. That view was diametrically opposed to what Cuvier was saying. One problem for Cuvier was that catastrophism came to be associated with the biblical Noachian flood, which had long been used to explain the presence of marine fossils found on land. In the Age of Enlightenment, however, any idea that was associated with religious dogma was automatically considered suspect.

But there was another factor involved. Cuvier's insistence that *this change was sudden, instantaneous not gradual* was made before anybody realized the true antiquity of the Earth. So when Cuvier saw the abrupt changes in fossil assemblages of the Paris Basin, it was perfectly logical to think of "abrupt" as being synonymous with "rapid". What he didn't realize is that the sharp contacts between two different beds of sediment could represent many millions of years when there simply was no sedimentation. But most of all, Cuvier's ideas suffered from the fact that he didn't have any clear answer to the question of what caused the catastrophes in the first place. Cuvier mentioned the possibility of "deluges" on the land, and seas drying up, but he didn't really know. And at that time he couldn't know. That had to wait till later.

Darwin's theory of evolution also contributed to the wide acceptance of uniformitarianism. Darwin, clearly influenced by Lyell, thought that it was far more likely that biological changes within species occurred slowly and continuously. That thinking paralleled the uniformitarianism model of Lyell. So, by the early 20th century the concept of uniformitarianism was firmly ensconced in both biological and geological thinking. Interestingly, early 20th century histories of the geosciences tended to view the catastrophism/uniformitarian controversy as a morality play that was ultimately won by the intellectually superior uniformitarianism. As the historian David Adams observed[1] "the publication of Lyell's *Principals of Geology* dealt catastrophism its death blow".

Catastrophism wasn't just out of fashion, it was wrong.

But Cuvier's early observation that fossil assemblages could change dramatically between successive layers of sedimentary rocks didn't go away. In fact, that observation was repeated as geologists learned to make maps of rock units. William Smith (1769-1839) used changes in fossil assemblages to unravel and map the stratigraphic succession of the sedimentary rocks underlying most of England.[2] But the underlying cause or causes of the abrupt fossil assembly changes remained a mystery.

If anything, the mystery deepened during the 20th century. In the 1950s, a paleontologist named Allison R. "Pete" Palmer was working on his Ph.D. dissertation cataloging the kinds of trilobites, an ancient and extinct group of three-lobed arthropods that were abundant in the Riley Formation of central Texas. As he sampled upward in the section, he noticed something very odd. Throughout most of the formation, the trilobite fauna was extremely diverse with anywhere from a dozen to more than 30 species coexisting at the same time. But near the top of the section, separated by just a few millimeters, the diversity ended abruptly. Instead of 20 or thirty trilobite species coexisting, now there were just one or two species present.[3] What could possibly have happened? It was Cuvier's *this change was sudden, instantaneous not gradual* observation all over again.

But the abrupt change from high to low species diversity was not all. As Palmer continued sampling upward in the section, the number of trilobite species began to increase again, and eventually the diversity approached the levels he had seen lower down. Then *bam* the diversity crashed again, once again over a very thin interval. And again you can just imagine Curvier's voice whispering *Living things without number were swept out of existence by catastrophes*. As Palmer and others continued to work, adding measured sections throughout the western United States, they eventually identified as many as seven packages— what Palmer termed biomeres[3]—of what for all the world looked like repeated and abrupt mass extinctions of trilobite species.[4]  But since this was in the 1960s and 70s, with Lyell's uniformitarianism deeply embedded in geological thinking, there wasn't much appetite for thinking in terms of catastrophic mass extinctions.

The uniformitarianism thinking, however, got a jolt in 1980 when Louis Alverez and his son Walter stunned the geological world by proposing that the end of the age of dinosaurs had been caused by a catastrophic asteroid impact.[5]  As often happens in science, the Alverezs came to this startling conclusion while looking at a totally unrelated problem.  One of the hardest things to do in sedimentology is to estimate the rate at which sediments accumulate.  Sediments can accumulate to thicknesses of hundreds of meters virtually instantaneously (terrestrial or submarine landslides), or to thicknesses of a few millimeters in millions of years (deep ocean seafloors).

It turns out that the element iridium is virtually absent in earth's crust.  That's because it is strongly siderophilic (iron loving).  Early in earth's history, when the planet was still molten, most of the siderophilic elements like iridium and uranium dissolved in the molten iron and migrated to the core of the earth, leaving very little in the earth's outer crust.  But iridium is abundant in the iron-nickel meteorites which regularly burn up in the earth's atmosphere at a predictable rate.  Walter Alverez, a geologist, got the idea of measuring iridium in marine sediments in order to calculate the rate that the sediments accumulated.  When he measured the amount of iridium in a clay layer spanning the period of time in which the dinosaurs became extinct, however, the concentrations of iridium were ten to a hundred times higher than normally found in the earth's crust.  That, in turn, suggested that an iron-nickel asteroid had hit the earth and vaporized, spreading a thin layer of iridium-laden dust over the whole planet.  It also happens that Louis Alverez was an atmospheric physicist. Louis showed mathematically that such an impact would create a huge dust cloud that could persist for decades, block out the sun, burn or kill most of the earth's vegetation, and cause a mass extinction that would be especially severe for large species of dinosaurs.

The Alvarez theory was greeted skeptically, to put it mildly, by the geological community. Part of that skepticism was just the normal vetting process that any new idea gets in the course of things. But a big part of it was that it raised the specter of Cuvier's catastrophism, which most geologists had been raised to believe was both antiquated and wrong. All of this was debated for years, with reports of iridium-rich rocks found at the dinosaur extinction boundary in several widely spaced places on earth. The clinching piece of evidence came when two geophysicists prospecting for oil in the Caribbean Sea found evidence for an impact crater located in the Yucatán Peninsula.[6] Later it was shown that the age of the impact crater coincided exactly with the demise of the dinosaurs, just as the Alvarez's had claimed.[7] Here, for the first time since Cuvier brought the issue up two centuries earlier, was indisputable evidence that catastrophic events could indeed cause the extinction of vast swaths of species.

Catastrophism had been rehabilitated.

All of this turned out to be a good thing for the geological community. As geologists had fanned out over the globe in the 19th and 20th centuries, they had found lots of evidence of the mass extinctions envisioned by Cuvier. Even before the Alvarez's proposed their asteroid hypothesis, there was firm paleontological evidence for at least 15 mass extinctions that had occurred between 540 million years ago to the present day.[8] In the 1980s, the open question was what had caused those extinction events? Could they all be traced back to asteroid impacts? Or could it be that other catastrophes like volcanic eruptions, earthquakes, global warming, global cooling, changes in ocean chemistry, changes in the magnetic polarization of the earth, or even bursts of cosmic radiation from space could also cause mass extinctions?

The short answer seems to be "all of the above". But it's hard to say for sure because it turns out to be extraordinarily difficult to track down the exact cause(s) of mass extinctions. The best example of this is the granddaddy of all mass extinctions, which occurred 251 million years ago. This extinction marks the end of the Permian Period of geologic time and thus is called the end-Permian extinction. It's difficult to even imagine how devastating the end-Permian extinction was to life on earth. As many as 95% of all marine species became extinct within a time frame of less than a million years.[9]

So what caused the end-Permian extinction event? Several investigators have claimed to have found evidence of an asteroid impact, but that evidence is far less conclusive than for the Alvarez's end-dinosaur extinction. There is, however, very good evidence of gigantic volcanic eruptions that occurred in Siberia at the same time as the extinction event. The term "gigantic" is no understatement. It is estimated that as much as two million cubic kilometers of basaltic lava was erupted, and the lava covered an area of 1,600,000 square kilometers[8] in what is now eastern Russia. This volcanic activity appears to have rapidly warmed the earth's climate, based on stable oxygen isotope evidence, which in turn caused the oceans shift from being fully oxygenated to being largely anoxic.[10] Without oxygen to breath, most of the species living in the world's oceans were asphyxiated and very quickly became extinct. But while the causes of the end-dinosaur and end-Permian extinctions are largely agreed upon, the reasons for the other 13 or so extinction events are far less clear.

All of these extinction events, regardless of the many, many things that might have caused them, do have something in common. While species diversity is (by definition) reduced by extinction events, in every instance species diversity immediately rebounded after the event. Stated another way, extinctions have the effect of emptying particular ecological niches. That, in turn, gives other species an "opening" to move into those niches.

So, the question is are extinction events uniformly a bad thing or can they actually be beneficial? Obviously becoming extinct *is* a disaster for any particular species. But a disastrous extinction for one species opens the door of opportunity for other species to step in and fill the vacated ecological niche. And that leaves us with an answer that is uniformly unsatisfying to everyone.

Yes and no.

-------------------------------

Several years after the young woman took her introductory ecology and geology classes at Montgomery College, and after she had transferred to the University of Maryland to finish her geology degree, she was fortunate enough get a student internship in paleontology at the Smithsonian Institute in Washington D.C. She was working for O.L. (Ollie) Karklins, an expert on marine bryozoan fossils[11], and John Pojeta an expert on mollusk fossils.[12] For years, Karklins and Pojeta had been collecting and cataloging fossil bryozoans and mollusks from a variety of locations throughout the world. In modern paleontology, the idea is not to collect just the "best" looking fossils that you can find, as most amateur fossil-hunters do. Rather, the idea is to take a carefully measured two or three-foot vertical section in an outcrop, break the rocks up with a sledge hammer, shovel all of the rock pieces into bags, and haul them back to the lab for analysis. That way, you are decreasing the sampling bias that comes from collecting just those fossils that happen to catch your eye. In the lab, the fossils are separated, identified, and sorted so that an estimate of relative species abundances can be made. That is a lot of work, and that's why Karklins and Pojeta needed student interns to help with it.

The young woman was fortunate to have the help of an experienced lab assistant named Maria. For the first week or so on the job, Maria patiently showed the young woman how to pick through the rocks, identify the different species present, and estimate their relative abundances. Once that was done, the young woman's next job was to collect a subsample weighing maybe ten or twenty pounds—out of the hundred pounds or more of rocks originally in the bags—that was "representative" of the whole. Lastly, she picked out individual specimens of each fossil type, cut them into sections with a rock saw, and made acetate peels for microscopic examination.

The collection that Karklins had her working had come from Kentucky and spanned the upper Cambrian (~475 million years) and into the lower Ordovician (~ 485 million years). The collections were stored in ascending drawers that were stacked in stratigraphic order from older to younger. She began with the older drawers and worked her way up the section. The drawers with Cambrian fossils consisted mainly of brachiopods and trilobites, and she had little trouble creating subsamples that were representative of the whole rock.

Finishing with one drawer, she opened the next one and froze in astonishment. Everything had changed. The kinds of brachiopods present were much different than the ones she had seen in the previous drawer, and there was a new kind of fossil present that she tentatively identified as a bryozoan. Hurriedly, she went to find Karklins to ask him about what she was seeing. When she did, Karklin's eyes twinkled with pleasure. She had just seen first-hand, as Karklins had slyly arranged, the boundary between the Cambrian and the Ordovician Periods of geologic time. That boundary marks a mass-extinction event in which as many as 25% of the previously existing marine genera disappeared. On the other hand, that boundary also marks the appearance of bryozoans, a new phylum of tiny sea creatures who lived communally in "apartment building" structures made out of calcium carbonate. Which again begs the question.

Was the end-Cambrian extinction event a catastrophe or an opportunity?

# REFERENCES

1.  Adams, F.D., 1938. The birth and development of the geological sciences. Dover Publications, Inc., New York, 505 pp.

2.  Winchester, S., 2001. The Map that Changed the World. William Smith and the Birth of Modern Geology.–1–332. HarperCollins Publishers, New York. 330 pp.

3. Palmer, A.R., 1965. Biomere: A new kind of biostratigraphic unit. Journal of Paleontology, pp.149-153.

4. Palmer, A.R., 1998. A proposed nomenclature for stages and series for the Cambrian of Laurentia. Canadian Journal of Earth sciences, 35(4), pp.323-328.

5. Alvarez, L.W., Alvarez, W., Asaro, F. and Michel, H.V., 1980. Extraterrestrial cause for the Cretaceous-Tertiary extinction. Science 208(4448):1095-1108.

6. Penfield, G.T. and Camargo, Z., A., 1981, Definition of a major igneous zone in the central Yucatán platform with aeromagnetics and gravity: Society of Exploration Geophysicists Technical Program. Abstracts and Biographies, 51, p.37.

7. Montanari, A., Margolis, S.V. and Claeys, P., 1992. Coeval 4" Ar/39Ar Ages of 65.0 Million Years Ago from Chicxulub Crater Melt Rock and Cretaceous-Tertiary Boundary Tektites. Science, 257, p.14.

8. Sepkoski, J.J., 1982. Mass extinctions in the Phanerozoic oceans: a review. Geological Society of America Special Papers, 190, pp.283-290.

9. Benton, M.J. and Twitchett, R.J., 2003. How to kill (almost) all life: the end-Permian extinction event. Trends in Ecology & Evolution, 18(7), pp.358-365.

10. Wignall, P.B. and Hallam, A., 1992. Anoxia as a cause of the Permian/Triassic mass extinction: facies evidence from northern Italy and the western United States. Palaeogeography, Palaeoclimatology, Palaeoecology, 93(1), pp.21-46.

11. Karklins, O.L., 1984. Trepostome and cystoporate bryozoans from the Lexington Limestone and the Clays Ferry

Formation (Middle and Upper Ordovician) of Kentucky. U.S, Geological Survey Professional Paper 1066-I

12. Pojeta Jr, J., 1988. The origin and Paleozoic diversification of solemyoid pelecypods. New Mexico Bureau of Mines and Mineral Resources Memoir, 44, pp.201-271

# CHAPTER 11
## TIME OUT OF MIND

After deciding that he wanted to major in geology, the young boy—now a young man—registered for Geology 102, Historical Geology. After Introductory Geology 100, Historical Geology was the next course in the sequence leading to a Bachelors degree. As the name implies, it focused mainly on a survey of the history of the earth. The professor was a paleontologist named Peter Stifel who, because of his gregarious nature and wry sense of humor, was very popular with the students. That popularity took a hit on the first day of class when he handed out copies of the Geologic Time Scale (Fig. 11.1) and announced that every student would need to memorize it by the first hour exam. The announcement was met by a chorus of groans by the students who, like students the world over, hated to memorize anything.

| Eon | Era | Period | Epoch | Age (MY) |
|---|---|---|---|---|
| Phanerozoic | Cenozoic | Quaternary | Holocene | |
| | | | Pleistocene | 1.5 |
| | | Neogene | Pliocene | |
| | | | Miocene | 23 |
| | | Paleogene (Tertiary) | Oligocene | |
| | | | Eocene | |
| | | | Paleocene | 65 |
| | Mesozoic | Cretaceous | | |
| | | Jurassic | | |
| | | Triassic | | 250 |
| | Paleozoic | Permian | | |
| | | carboniferous Pennsylvanian | | |
| | | carboniferous Mississippian | | |
| | | Devonian | | |
| | | Silurian | | |
| | | Ordovician | | |
| | | Cambrian | | 541 |
| Precambrian | | | | > 541 |

Figure 11.1—A simplified Geologic Time Scale.

"No bitching" Dr. Stifel replied sternly, "the geologic time scale is something you need to know and love".

So, the young man dutifully sat down and memorized the geologic time scale over the next couple of weeks. A week or so before the first hour exam, Dr. Stifel gave a pop quiz at the beginning of class. The quiz consisted of a single blank piece of paper upon which the time scale was to be reproduced. The point of the quiz, which didn't count for a real grade, was simply to help the students see how their memorization was proceeding. Accordingly, there were only two grades he gave. A "zero" if there was even one mistake, and a "hundred" if there were no mistakes. When he got his paper back the next class, the young man was mortified to find he had gotten a zero. It turns out he failed to capitalize "Pleistocene", a grievous error indeed because, as every geologist is supposed to know, the names of the epochs are *always* capitalized.

The significance of the geologic time scale, however, is much more than just a convenient way to torment undergraduates. It is a contrivance that allows the human mind to comprehend the incomprehensible enormity that is geologic time.

------------------------------

The first attempts to determine the age of the earth were based on the genealogies found in the Bible. In the 17th century, a Bishop James Usshur (1581-1656) famously calculated that the time and date of creation was "the entrance of the night preceding the 23rd day of October... the year before Christ, 4004". That is, the earth was created at 6 pm on the 22 of October, 4004 BC. That date, interestingly enough, is not that much different from other estimates calculated from Bible genealogies by such luminaries as the Venerable Bede (3952 BC), Johannes Kepler (3992 BC), and Sir Isaac Newton (ca. 4000 BC). The differences in these estimates are due primarily to assumptions that had to be made about the length of a human lifespan. We might be tempted to look on those estimates with a bit of a smile, but they illustrate an important point. It is perfectly natural for people to think of time in reference to the normal human lifespan. That measure of time works perfectly well for almost all human-based needs.

But it doesn't work for geologic time.

The geologic time scale that we use today (Fig. 11.1) was the result of a very improbable sequence of very strange events that happened over three full centuries. The first of these improbable events involved a Danish physician and a shark's tooth. Niels Stensen, a native of Copenhagen, had left Denmark in 1660 to study medicine at the University of Leiden in the Netherlands where, in due course, he specialized in human anatomy. When he completed his studies, he moved to Florence, Italy in 1665, procured employment at a hospital, and continued his studies of the human musculature under the patronage of the Duke of Tuscany.

That's when things began to get strange.

In 1666, a couple of Italian fisherman happened to catch a huge shark, an unusual occurrence that immediately drew a lot of local interest. Intrigued, the Duke of Tuscany sent the shark's head to Stensen, who was now using the latinized name Nicolaus Steno, to be dissected and studied. Steno dutifully performed the dissection for his patron, carefully drawing the head with its jaws open. With equal care, Steno drew each individual tooth in order to document the shark's impressive rows of teeth found therein. As he examined the shark's teeth, he couldn't help noticing their resemblance to the enigmatic three-sided objects called "tongue stones" found in the sedimentary rocks of Tuscany. These objects got that name, presumably, because they vaguely resembled a human tongue (Figure 11.2).

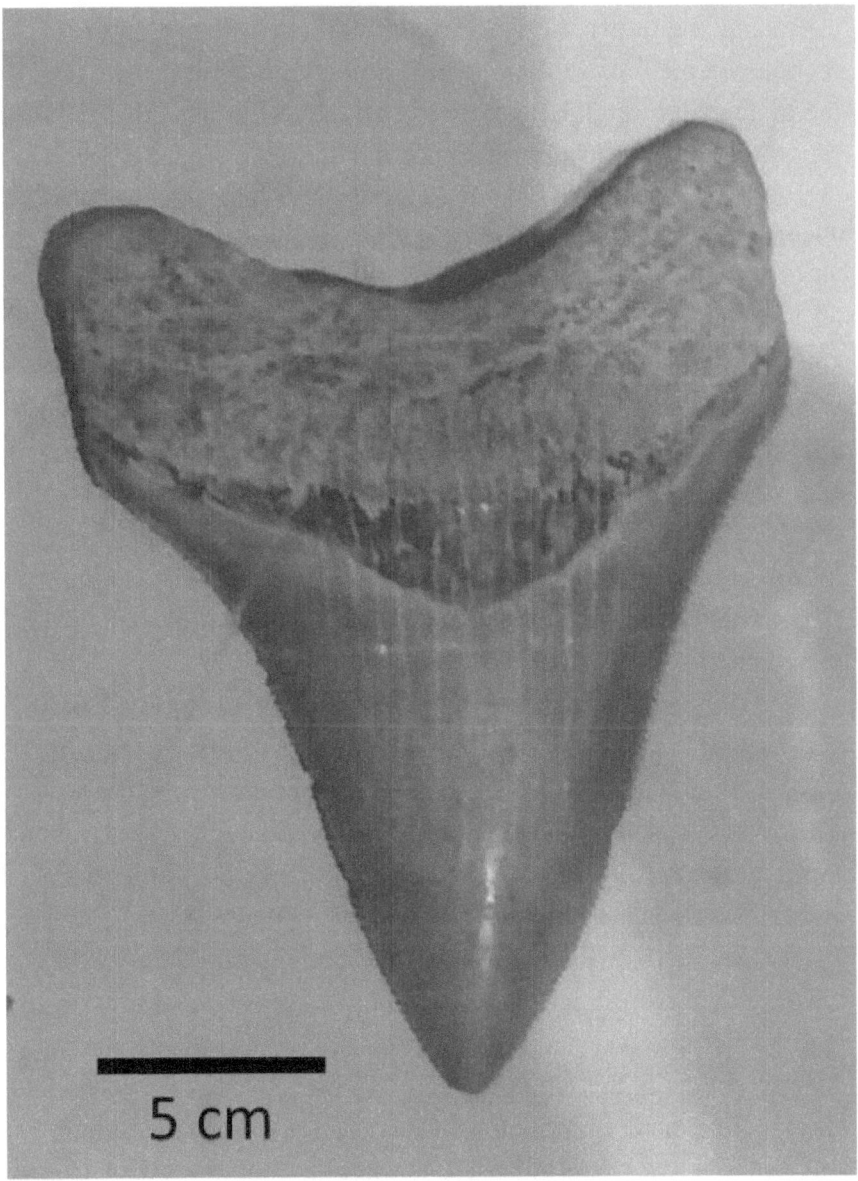

5 cm

Figure 11.2—A three-sided sharks tooth (*Carcharodon megalodon*) or a "tongue stone".

In the first century AD, the Roman naturalist Pliny the Elder had noticed the tongue stones and, puzzled, wondered if they might have fallen from the sky. But Steno, by virtue of having examined and drawn the shark's head and teeth, immediately recognized the tongue stones for what they really were—sharks teeth. Furthermore, Steno realized that they must have been deposited in the sea since, after all, that's the only place where sharks live. Steno published his ideas in 1669 in a treatise named *Preliminary discourse to a dissertation on a solid body (*i.e. a shark's tooth*) naturally contained within a solid* (i.e. a sedimentary rock). Steno reasoned that the teeth must have fallen from the sharks mouth and been deposited on the bottom of the sea along with other sediments. He also reasoned that, since sediments were continually being deposited on the seafloor, more deeply buried sediments must be older than shallower sediments. Thus was born Steno's law of superposition: *layers of sedimentary rocks represent a **time sequence**, with the oldest at the bottom and the youngest at the top.* In other words, the concept of *time* could now be applied to sedimentary rocks.

It's hard to imagine just how revolutionary—and alien— that notion was in 1669. It's perfectly natural to think of time in the dynamic context of days, years, or human lifespans. But to think of time as being recorded by *static layers of rock* was a new thing under the sun. Given that Steno published his findings in Italy, it's almost certainly no coincidence that another Italian named Giovanni Arduino (1714-1795) made the first attempt to classify *time* according to the succession of geologic strata.

Arduino was a mining geologist who became intrigued by the succession of metamorphic and sedimentary rocks exposed on the southern flank of the Italian Alps. The core of the mountains, he observed, consisted of hard metamorphic rocks that were clearly overlain by lithified sedimentary rocks on the flanks of the mountains. These lithified sedimentary rocks, in turn, were overlain by unlithified sediments. Using Steno's principal of superposition, Arduino named the metamorphic rocks *Primary* since they were at the bottom of the sequence and therefore the oldest. Similarly, he named the lithified sedimentary rocks

*Secondary* because they were younger than the Primary rocks. Next he named the unlithified sediments *Tertiary* because they were younger still. Finally, and because Italy is volcanically active, volcanic rocks were observed to overlie Tertiary and Arduino called them *Volcanics*. Arduino published this first attempt at a geologic time scale in 1759. Its main significance was that it was it took Steno's principal of superposition to its next logical step, which was to stack geologic sequences in terms of their relative ages (Fig. 11.3).

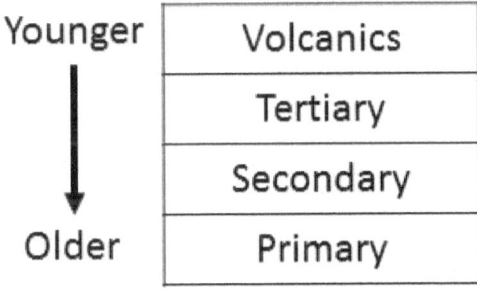

Figure 11.3—Giovanni Arduino's geologic time scale (1759) based on rocks exposed in the Southern Alps of Italy, with relative ages assigned using Steno's Principal of Superposition.

The next strange occurrence in the development of the geologic time scale had to do with, of all things, fashion. For whatever reason, sometime in the early 18th century it became fashionable for well-off English gentry to collect and exhibit fossils in their homes[1]. There doesn't seem to be any particular reason for this other than it was also fashionable for the gentry, including both men and women, to be amateur naturalists. As such they made extensive collections of all sorts of things including butterflies, birds, beetles, and......fossils. Fossils seem to have been particularly popular objects for collecting, partly no doubt because well-preserved fossils can be rather beautiful. So, throughout much of the 18th century extensive collections were made of the fossils extracted from the sedimentary rocks that underlie much of England.

It was a surveyor named William Smith (1769-1839) who first came to the realization that fossils were more than just pretty bobbles. Smith, who had been interested in fossils since boyhood, was lucky enough to get a job surveying a new canal in Somerset. Somerset was coal country and canals were needed to move coal from the mines to the industrial centers of Bristol, Sheffield, and London. During construction of the Somerset canals, he carefully documented how the morphology of different marine fossils changed as the excavations revealed progressively older layers of strata. Given Steno's law of superposition, Smith gradually came to the realization that the fossils were also *a record* of *the passage of time*. The significance of that realization astonished Smith, who in 1796 wrote[1] that:

> *Fossils have long been studied as great curiosities,*
> *collected with great pains, treasured with great*
> *care and at a great expense, and showed and*
> *admired ...... because it is pretty; and this has been*
> *done by **thousands who have never paid the least**
> ***regard to that wonderful order*** (my emphasis) *and*
> *regularity with which Nature has disposed of those*
> *singular productions, and assigned to each class (of*
> *fossils) its particular stratum.*

Of course, it being just 1796, Smith had no earthly idea *why* fossil morphology changed as you went from older to younger strata. Darwin's theory of evolution wouldn't be published for another 62 years. But, for the first time in human history, Smith had unequivocal evidence that fossils were unique to their position in the stratigraphic column, and that those changes reflected the passage of time. That astonishing insight made two things possible. First, it was now possible to correlate strata not just by their lithology (limestone, claystone, sandstone), but also by the fossils contained therein. Secondly, and probably more importantly, it made it possible to think of time not just in days, years, or lifespans, but in the span of time it takes for fossils to change from one form to another. What that time span might be, however, was a complete mystery and would remain so for another hundred and fifty years.

This brings up the next odd occurrence, which is probably better described as an incredible coincidence. By sheer fortuitous happenstance, the landmass of England is underlain by sedimentary rocks deposited over more than 600 million years. In addition, and again by sheer fortuitous happenstance, the strata all have been tilted so that they dip gently to the southeast. This means that the rocks are exposed at land surface from the youngest in the Thames River Valley to the in the oldest in the Welsh mountains. That, in turn, gave Smith and other British geologists a bird's eye view of much of the Earth's history. When Smith published his geologic map of England in 1815[1], he was careful to illustrate this stratigraphic succession by drawing a cross section showing how the geologic formations are sequentially exposed from east to west (Figure 11.4).

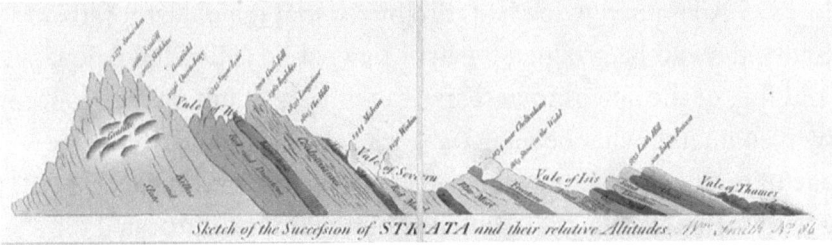

Figure 11.4. William Smith's drawing showing in cross section the rocks ranging in age from the youngest in the Thames Valley to the oldest in Wales.

Notice that Smith described each layer of sedimentary rock simply as "limestone", "red marl", or "coal measures", not by formational names. That's because in 1815 the practice of formally naming rock formations had barely begun. But, given the publication of Smith's map, it didn't take geologists long to begin that process.

The final odd occurrence leading to the geologic time scale had to do with William Smith's sister. She had married a Welshman named Philips, and they had a son named John Philips. But unfortunately, both parents died when the boy was just seven years old. William Smith, being the boy's uncle, took him into his own family and saw to John's care and education. When John was fifteen years old, he became Smith's surveying assistant, and began accompanying Smith on his trips around the country as an itinerant surveyor. As you might expect, John Philips absorbed much of Smith's knowledge about fossils and how those fossils could be used to correlate geologic strata. Having that knowledge proved to be a good career move for John, who at the age of 24 secured a position with the Yorkshire Philosophical Society organizing and classifying the Yorkshire Museum's extensive fossil collections.

As Phillips sorted through the vast fossil collections now at his disposal, he came to the realization that the fossil record of England could be divided into three broad subdivisions that were separated by what appeared to be catastrophic extinction events. In 1835 Adam Sedgwick had recognized that the oldest fossils in England could be grouped together in what he called the Paleozoic (old life, or the age of invertebrates). Phillips built on that concept by recognizing what he named the Mesozoic (middle life, or the age of reptiles), and the youngest the Cenozoic (new life, or the age of mammals) in 1840. Interestingly, Phillips illustrated the extinction events separating these subdivisions by showing the relative number of fossil species on the horizontal axis compared to the passage of time (on the vertical axis) (Figure 11.5).

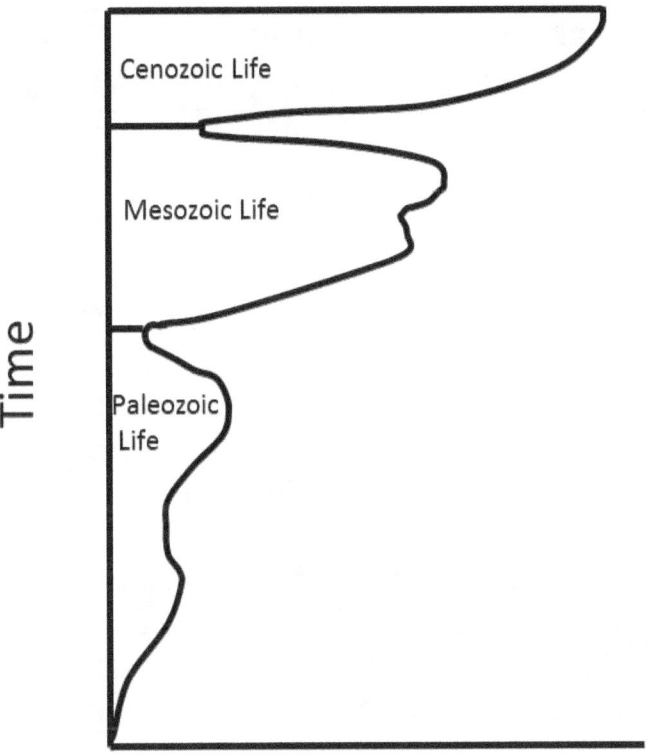

Cenozoic Life

Mesozoic Life

Paleozoic Life

Time

## Relative number of species

Figure 11.5—The geologic time scale created by John Phillips (1840) in which the decrease in the number of fossil species, or extinction events, conveniently separated geologic time into the Paleozoic (old life), Mesozoic (middle life) and the Cenozoic (new life).

Throughout the 19th century, numerous geologists contributed to the construction of what became the geologic time scale by recognizing and naming subdivisions of Paleozoic, Mesozoic, and Cenozoic time. A list of some prominent contributors to this process, and the publishing dates of those contributions, is shown in Table 11.1 [2]

| Time period name | Time period namer | Year named |
|---|---|---|
| Primary | Arduino | 1759 |
| Secondary | Arduino | 1759 |
| Tertiary | Arduino | 1759 |
| Volcanic | Arduino | 1759 |
| Jurassic | von Humbolt | 1795 |
| Triassic | von Alberti | 1815 |
| Coal Measures | Farey | 1807 |
| Cretaceous | d'Omalius d'Halloy | 1822 |
| Carboniferous | Conybeare & W. Phillips | 1822 |
| Quaternary | Desnoyers | 1829 |
| Eocene | Lyell | 1833 |
| Cambrian | Sedgewick | 1834 |
| Paleozoic | Sedgwick | 1835 |
| Silurian | Muchison | 1835 |
| Recent | Lyell | 1839 |
| Pleistocene | Lyell | 1839 |
| Pliocene | Lyell | 1839 |
| Miocene | Lyell | 1839 |
| Devonian | Murchison & Sedgewick | 1840 |
| Mesozoic | Philips | 1840 |
| Permian | Murchison | 1841 |
| Cenozoic | Philips | 1844 |
| Oligocene | von Beyrich | 1854 |
| Paleocene | Schimper | 1854 |
| Neogene | Hömes | 1859 |
| Precambrian | Salter | 1864 |
| Paleogene | Neumann | 1866 |
| Archaean | Dana | 1872 |

| Ordovician | Lapworth | 1879 |
|------------|----------|------|
| Holocene | Gervais | 1885 |
| Proterozoic | Emmons | 1887 |
| Hadean | Cloud | 1972 |
| Pleistogene | Harland | 1989 |

Table 11.1—Chronology of the adoption of geologic time names.

As the geologic time scale developed, it became increasingly obvious that, in order for all of these sedimentary rocks to be deposited, as well as for the different life forms to develop, become extinct, and then to be replaced by other species, an enormous amount of time must be involved. The problem was, while the relative ages of the different subdivisions had been worked out by about 1880, there was no way to know their absolute ages. In 1860, just a year after the publication of his *Origin of Species*, Darwin somewhat haphazardly guessed that the age of the earth to be about 300 million years old. John Phillips immediately took Darwin's "abuse of arithmetic" to task and subsequently came up with his own estimate of 100 million years. But those estimates were essentially guesses based on assumptions that were impossible to verify or refute. And so the mystery of the Earth's real age persisted.

It wasn't until 1896, when the French physicist Henri Becquerel noticed that uranium-bearing minerals were emitting x-rays, a phenomenon that Pierre and Marie Curie later named radioactivity. The Curies went on to purify radium, another radioactive element. When radium decays it produces the nucleus of a helium atom stripped of its electrons which Ernest Rutherford originally named an alpha particle. In 1905, the British physicist John William Strutt got the idea of measuring the amount of helium present in a radium-containing rock. Having an approximation of the radium's half-life (1,600 years), Strutt estimated the rock had to be more than a billion years old. The era of radioactive dating had begun.

It was the American Chemist Bertram Borden Boltwood (1870-1927) who noted that lead was always present in uranium ores and correctly reasoned that lead was one of the breakdown products of uranium decay. In 1907, at the suggestion of Ernest Rutherford, Boltwood measured the amount of lead in a uranium ore, and knowing the approximate half-life of uranium 235, he calculated that the rock was 2.2 billion years old. Boltwood never bothered to publish his results, but an undergraduate studying geology at Imperial College in London named Arthur Holmes heard about it and immediately realized its potential to date rocks. Improving on Boltwood's methods, Holmes was able to date rocks from a Devonian sequence in Norway as being 370 million years old. That was the first radiometric date for a rock that had ever been made. Furthermore, the presently accepted dates of the Devonian Period (419.2 to 358.9 million years before present) indicate that even by modern standards Holmes' age date was fairly accurate.

Holmes published his results in 1911 just after he graduated from college. Three years later he published a book entitled *The Age of the Earth* in which, in addition to his Devonian date (370 MY) he provided a date of 240 MY for the Carboniferous and 430 MY for Silurian/Ordovician rocks. The book's main point, however, was simply to advocate radiometric dating as a way to assign absolute ages to the relative ages given by the geologic time scale as it had developed in the 19th century.

The twin cataclysms of World War I and World War II combined to slow progress on radiometric dating techniques between 1910 and 1945. But ironically, interest and funding for studying uranium and its daughter products generated by the Manhattan Project helped improve analytical methods following WW II. In 1947, and with the help of a former Manhattan Project chemist named Alfred Nier, Holmes was able to assign absolute ages to five samples collected from known positions in the stratigraphic column, and produce a geologic time scale[3] that, for the first time, was based on actual age measurements.

Since 1947, the absolute ages associated with the geologic time scale has been progressively refined and improved (Fig. 11.1). Given the precision of modern radiometric dates, you might think that traditional names for the geologic Eras, Periods, and Epochs (Paleozoic, Cambrian, Pleistocene, etc) would fall out of use. After all, now that we can now reliably date the actual ages of most rock sequences, why not simply refer to them by their actual ages (i.e. 360, 240, 56 MY)? But geologists don't generally do that. Rather, they still tend to refer to rocks as being "Cambrian in age" rather than "530 MY old". Part of this, no doubt, is the two hundred year habit of equating particular rock sequences and their fossil assemblages in terms of the traditional age names. But habit probably doesn't entirely explain it.

Human beings, by virtue of our own biology, are naturally constrained to think in terms of the number of minutes we have to get to work, the number of hours of sleep we got last night, the number of months in a pregnancy, or the number of years in a lifetime. Our brains, therefore, simply aren't naturally equipped to imagine time spans involving hundreds of millions of years. We can, however, associate "Cambrian" time with a mental images of the swarms of trilobites that swam in the seas 500 million years ago. We can also associate coal seams with the idea of a "Carboniferous" period of time 350 MY years ago, or a hulking Tyrannosaurus Rex with "Cretaceous" time 70 MY ago. Even relatively recent times of, say, 50,000 years ago are hard for people to get their minds around. It's easier to conjure up images of newly evolved, spear-toting, bearded *Homo sapiens sapiens* chasing wooly mammoths during "Pleistocene" time. The modern geologic time scale is not just a tabulation of chronology. It's a coping device. It enables our brains to deal with the incomprehensible expanses of time that are earth's history.

-------------------------

It took a couple of weeks, but eventually all of the students in the Geology 102 class, got 100s on the quiz reproducing the geologic time scale. The young man accomplished this on his second try. Dr. Stifel, from long practice and habit would tacitly acknowledge the tens and hundreds of millions of years he was lecturing about, but he always did it by speaking in terms of Triassic, or Cretaceous, or Oligocene time. To do it any other way—associating specific happenings in earth history to numerical ages—is just too difficult to keep straight.

Geologists aren't the only people who use this kind of time shorthand. Egyptologists, for example, think in terms of the "Old Kingdom" rather than saying 2,686 – 2,181 BC. The Old Kingdom brings to mind the construction of the great pyramids of Giza, something the human brain can actually relate to. In the same way, rather than saying 1,550-1077 BC they refer to the "New Kingdom" which immediately brings to mind the treasures of Tutankhamun's tomb. It's a mental shorthand that just makes it easier to remember and organize a lot of information.

It's just how humans are wired.

## REFERENCES

1. Winchester, S., 2001. The Map that Changed the World. William Smith and the Birth of Modern Geology.–1–332. HarperCollins Publishers, New York. 330 pp.
2. Hay, W.W., 2016. Experimenting on a Small Planet: A History of Scientific Discoveries, a Future of Climate Change and Global Warming. Springer, 818 pp.
3. Holmes, A. 1947. The construction of a geological time-scale. Transactions of the Geological Society of Glasgow, 21, 117-152.

# CHAPTER 12.
# A MEETING IN MINERALOGY

With considerable trepidation, the young woman walked into in the mineralogy laboratory classroom of the Geology Department. She had just transferred from Montgomery College, a relatively small cozy campus, to the sprawling College Park campus of the University of Maryland. The sheer size of the place was intimidating and at the time (1973) it accommodated more than 20,000 undergraduates. The Geology Department itself was relatively small with just five or six professors, was only a couple of years old, and had no graduate program. Mineralogy was one of the "core" subjects that all geology majors had to take and the young woman was apprehensive about how she would do.

The 1970s were a time of explosive growth for departments of geology in American universities. The excitement surrounding plate tectonics alone would no doubt have increased student enrollment in geology programs. But then came the Arab oil embargo of 1973 which quadrupled the world price of oil in less than a year. That sparked an explosion of exploration and drilling for oil in the United States and other parts of the non-Arab world. That explosion generated a huge need for trained geologists to work in oil exploration and development. That led to dramatic increases in geologist salaries, which naturally led to large increases in enrollments in geology programs around the country. The University of Maryland was no exception. Since the Geology Department had been founded just three years earlier, enrollment had increased rapidly from a handful to more than a hundred geology majors. The young woman, who had just transferred in from Montgomery College, was one of them.

The young man, also a declared geology major, had already taken a seat at a bench when the young woman walked into the mineralogy lab. He noticed her immediately. She was tall and slender, with center-parted soft brown hair that fell almost to her waist. Careful to avoid staring, the young man followed her out of the corner of his eyes as she selected a place at another bench and sat down. Wow, he thought.

A beauty.

--------------------------

You might think that there's not much about mineralogy—the study of the physical and chemical properties of minerals—that would even remotely qualify as being romantic. But curiously enough, beginning at least with the ancient Greeks, it was widely thought that minerals found in the earth were the product of—sex.

The Peripatetic School, which Aristotle founded in Athens in the fourth century BC, produced many talented and prolific pupils. One of these was a native of Lesbos named Theophrastus, who eventually became Aristotle's successor as the leader of the school. In his thirty-six years leading the Peripatetic School, Theophrastus produced numerous books about various topics in natural history. One of these, entitled *Concerning Stones*, is notable because it became the standard mineralogy textbook in Europe for the next 1,800 years. In this relatively short book, which may have originally just been lecture notes[1], Theophrastus describes various minerals and precious stones including emeralds, amethyst, onyx, and jasper.

But the really curious thing about Theophrastus' book is that he frequently describes minerals as being male or female. For example, in describing what he calls "carnelians" (chalcedony), he writes[1]:

> *That species which is pellucid and of a brighter red is called the female and that which is pellucid and of a deeper red with some tendency to blackness, the male. The Lapis Lyncurious* (zircon) *is distinguished in like manner, the female being more transparent and of a paler yellow; and the Lapis Cyanus* (lazurite) *is in the same manner divided into Male and Female specimens: the Male being deeper*

*in color.*

It's not clear whether Theophrastus considered some colors or crystal habits, by virtue of being rustic or stout as being "male", or others by virtue of seeming more delicate "female". Furthermore, it's not clear if he believed that the union of the two sexes could produce mineral "children" of the same species. What is clear, however, is that because Theophrastus' book was so widely copied and circulated for the next two millennia, some later writers clearly took him literally, believing that minerals could be produced by sexual union between the male and female crystals.

One Sir John Mandeville, for example, who purportedly traveled to India around 1350 AD, reported that male and female diamonds did indeed mate to produce new diamonds. Another example, which may just be a retelling of Mandeville's tale, is a book concerning gems written in 1556 by one Franciscus Rueus who relates the following story:

> *It was told me many years ago by a certain lady whose word could be trusted, that there was a Lady "Heurensis" descended from a celebrated Luxemburg family who had two diamonds which she had inherited and carefully preserved, which frequently by a miracle of nature produced other diamonds and that anyone who watched them at certain times would see that they passed through birth throes and produced offspring similar to themselves........*

All of this makes a little more sense when you consider that in the Middle Ages and into the Renaissance, scholarly consensus held that all matter could be classified as being animal, vegetable or mineral. And since both animals and vegetables were observed to reproduce sexually, it's not that great a leap of faith to think that minerals were also produced by sex. As a physician named Thomas Shirley wrote in 1672[1] when considering the origin of kidney stones:

> *As Vegetables and Animals have their Original*
> *from an invisible Seminal Spirit, or breath: so also*
> *have Mineral, Metals, and Stones.........*

Shirley goes on to explain this "Petrific Seed" theory for the origin of minerals:

> *That is, that the matter of all Bodies is originally*
> *meer water; which by the power of these seeds is*
> *coagulated, condensed, and brought into various*
> *forms.....*

In other words, water contains mineral matter, which if exposed to the proper "seed" would generate new minerals. As it happens, if you leave out the implication of sexuality, that's not too far from the truth.

----------------------------

The mineralogy professor, Dr. Galt Siegrist, was a slight man with a scruffy beard and a perpetual twinkle in his eye. He began the class by declaring that mineralogy, by his own design, was going to be the hardest class the students had ever had. So, he continued, you'd better listen up, take detailed notes, and expect to put in several nights a week doing the labs. He looked around the class, clearly enjoying the discomfort these words elicited from the twenty or so students in the class. Dr. Siegrist wasn't known as being a mean or cranky professor. On the contrary, he was usually quite cheerful and helpful, and he was a popular academic advisor. But he took mineralogy very seriously, and considered it to be one of the most important courses in the geology curriculum. As such, he was determined to make it as challenging—and thus as useful—as he possibly could.

This was going to be a long, hard semester for everybody.

The mineralogy class Dr. Siegrist had designed for the students was distilled from information accumulated over the two thousand years since Theophrastus wrote *Concerning Stones*. If you don't count his curious references to male and female gems and minerals, Theophrastus actually made some good contributions to what would become the science of mineralogy. For one thing, he made a stab at classifying minerals according to their physical properties. Theophrastus had noted that upon heating, some minerals either melt or change form in some way. The mineral cinnabar (mercuric sulfide or HgS) for example, when heated in air will give off the sulfur atom as sulfur dioxide and the mercury as a vapor which can then be condensed into liquid mercury. The ancient Greeks, who used metallic mercury to extract gold and silver from ores, were well acquainted with how heat changed cinnabar. Other minerals, particularly precious gemstones such as diamonds, are not noticeably transformed by heating. To Theophrastus, therefore, reaction to heat seemed a logical way of classifying different minerals. Much of Theophrastus' information seems to come from miners or quarryman who had a practical knowledge of the sources and uses of different minerals. That's probably one reason his treatise was considered to be so useful for so long.

The Romans also had a practical knowledge of mineralogy, mostly as it applied to ores of copper, tin, and iron. The Roman naturalist Pliny the Elder, arguably the western world's first encyclopedist, compiled his *Natural History* in the first century AD. In 37 books, Pliny laid out much of what was known (or thought to be known) at the time concerning the natural world. The topics Pliny considered included the kinds of animals, plants, gardening, medicine, metallurgy, and, in the $36^{th}$ and $37^{th}$ books, rocks and minerals. Some of his information, such as reporting that amber is the hardened gum of trees, is accurate. On the other hand he is also prone to repeating what are clearly old wives tales. For example, in describing what are possibly fluorospar crystals, for example, he remarks[1]:

> Which we may make bold to state is nothing but a
> frozen water, a very hard variety of ice, found in
> northern countries or on high mountains.

He does give credible descriptions of many gemstones including beryl, garnet, ruby, and spinel, and adds discussions of how they are used in jewelry. Pliny's *Natural History* was widely copied and circulated in the ancient world and appeared in print in 1469.

During the Middle Ages, much of what was known about rocks and minerals was derived largely from Theophrastus and Pliny the Elder. This information was compiled in various encyclopedias or in books about rocks and minerals called "lapidaries".[1] One interesting thing about these lapidaries is what they reveal about the medieval mind. The books rarely included original information collected by the authors, but rather tended to repeat what had been derived from other "scholarly" sources. The goal of many of these writers seem to be simply to repeat what had been said previously. On the other hand, when "original" information is given, it was generally more fanciful than useful.

One lapidary, for example, written in the 11th century by a man named Marbodus, is essentially a compilation of the purported magical or medical properties of minerals[1]. Many of the mineral names Marbodus uses are indecipherable. But the listed properties of those minerals that are recognizable can only be described as being weirdly medieval. For example, what Marbodus calls "adamas" (diamonds) "dispels evil dreams and saves its wearer from the influence of poison". Or what Marbodus calls "smaragdus", or green stones such as emerald or malachite, "enriches its wearer and makes him eloquent, wards off epilepsy and has many other virtues." To a man, the medieval writers of lapidaries are oddly unconcerned about the physical properties of minerals. Indeed, one gets the impression that many of the writers had never seen many of the minerals they are describing.

Volumes have been written about the fundamental change in human thinking that occurred between the Middle Ages and the Renaissance. A good illustration of this change, interestingly enough, is in the theory and practice of mineralogy. Throughout ancient times and into the Middle Ages, mineralogy was part practical identification of useful gems, ores, and minerals and part pseudo-magic. That approach came to a fairly abrupt end due largely to the discovery of rich deposits of tin and silver ores near Annaberg, Saxony in the 15$^{th}$ century. A thriving mining industry soon grew up and with it a renewed interest in practical mineralogy.

Georgius Agricola 1494-1555) was born with the given name of Georg Bauer (Bauer means "farmer" in German). He was initially trained as a classical scholar and, as was the custom in those days, he changed his name to the Latin equivalent (Agricola) of "farmer". He eventually gravitated to Italy where he studied medicine at the universities of Bologna and Padua. On returning to Saxony, he became the "town doctor" for the prosperous mining town of Chemnitz. As such, he was brought into continuous and intimate contact with the mining industry, and inevitably with the minerals upon which that industry was based. A good illustration of Agricola's firmly renaissance thinking is his opinion of the purported medicinal properties of minerals, upon which he commented[1]:

> *Of the powers which the Persian magicians and the Arabians attribute to certain stone and gems I will say nothing. Dignity and propriety obliges a man of science to reject them entirely.*

Agricola's lasting scientific contribution was to devise a classification system for minerals based solely on their physical properties such as color, weight, transparency, luster, taste, odor, shape and texture. His system of mineralogy, while a vast improvement on Theophrastus and Pliny, was inevitably hampered by the fact that he had no way of knowing the actual composition of the minerals. Agricola gave detailed descriptions of many new minerals, particularly metal ores. Of equal significance was his recognition that the mode of mineral occurrence (in veins, finely disseminated in rock, or crystal masses, etc.) and the associations between different minerals were important. All of this can be traced directly to his intimate knowledge and observations made in the mines of Saxony.

An important contribution to mineralogy was also made by Nicolas Steno (1638-1686). In the same book in which he formulated his Law of Superposition for sedimentary rocks[2], Steno also made the observation that the angles between the faces of crystals are the same for all specimens of the same mineral. That observation, now known as *Steno's Law of constant angles* or the *First law of crystallography,* became the basis for using crystal form to identify minerals.

Agricola and Steno's practical approaches to mineralogy were continued by Abraham Gottlob Werner (1749-1817), another Saxon whose father had been a mine inspector, and was thus intimately acquainted with the occurrence and associations of different minerals. Werner studied at the Mining Academy at Freiberg, and beginning with weekend excursions to different mines, began a mineral collection that eventually became the best in Europe. He became a lecturer at the Freiberg Mining Academy at the age of twenty-five where he taught for the rest of his life. He proved to be an inspiring teacher, and produced an entire generation of enthusiastic students who fanned out over Europe burning with geologic zeal. He also published, posthumously as it turned out, a system of mineralogy in which he divided minerals into classes based on their physical properties. These were:

*Class I.* Earthy minerals including silicates, clays, and talcs
*Class II.* Saline minerals including carbonates, nitrates, chlorides and sulfates.
*Class III.* Combustible minerals including sulfur, coals, bitumen, and graphite.
*Class IV.* Metallic minerals, gold, silver, copper etc.

All in all, Werner included 317 mineral in his list, which was beginning to resemble the classification schemes we use today.

It was the Swedish chemist Jöns Jacob Berzelius (1779-1848) who introduced chemistry into the study of mineralogy. Beginning in 1808, Berzelius began probing the composition of various inorganic substances. He measured the atomic weights of different elements and deduced the formulas of various oxides, sulfides, and salts. In 1813, a friend of his happened to give him a mineral collection, and Berzelius proceeded to demonstrate that the minerals Werner had classified as "earthy minerals" were in fact complex combinations of silica oxides and hydroxides. Thus, in addition to physical properties such as hardness, color, and crystal form, minerals could also be classified and identified based on their chemical composition.

The discovery of x-rays by Wilhelm Röntgen in 1895 set the stage for the next important development in mineralogy. As it happens, the wavelength of x-rays (1-100 angstroms) are similar to the spacing of the layers of atoms that make up of many minerals. Because of their regular atomic structure, crystals diffract x-rays according to the equation $2dsin\theta=n\lambda$ (Bragg's Law) where $d$ is the spacing between diffracting planes, $\theta$ is the incident angle of the x-rays, $n$ is any integer, and $\lambda$ is the wavelength of the x-ray beam. As early as 1912 the German physicists Paul Peter Ewald and Max von Laue postulated that x-ray diffraction patterns could be used to deduce mineral crystal structures. In the 1920s, a young Linus Pauling used x-ray diffraction to deduce the five *Pauling's rules* which summarized how the atoms arrange themselves in crystals depending on their relative sizes and electrical properties.

------------------------------------

The syllabus that Dr. Siegrist had developed over the years for his mineralogy course was a direct reflection of this history. The class began with an overview of crystallography, how atoms or molecules with a uniform chemical composition pack together to form crystals. Depending on how they fit together, the resulting crystals might have a cubic form (halite, or ordinary table salt), a hexagonal form (snowflakes, or frozen water), or rhombohedral form (calcite in seashells). Those crystal forms, in turn, could be used to identify different minerals. Next, the class would systematically work through the different chemical classes of minerals beginning with native elements (copper, sulfur), sulfide minerals (pyrite or "fool's gold"), oxides (hematite or iron oxide), halides (halite or table salt), carbonates (calcite as in sea shells), silicates (compounds of oxygen and silica such as quartz), and several other more exotic chemical compounds. Everybody in the class would be expected to memorize their physical properties (crystal form, hardness, luster, etc), their chemical formulas ($SiO_2$ for quartz, $KAl_2(AlSi_3O_{10})(OH)_2$ for muscovite, a common mica), and be able to identify hand specimens of each mineral on sight. Altogether, each student would end up with index cards summarizing all of this information, which they would commit to memory, for about 200 minerals.

After going over the syllabus and assigning the readings, Dr. Siegrist dismissed the class. As he was leaving, the young man again noticed the young woman with the long brown, center-parted brown hair. In the class of 20 students there were only three women, and so it was hard not to notice her. He wondered if he'd get a chance to meet her as the semester progressed.

As it happened, that meeting occurred later that very same day.

## REFERENCES

1. Adams, F.D., 1938. The birth and development of the geological sciences. Dover Publications, Inc., New York, 505 pp.

2. Steno, Nicolas, 1669. Preliminary discourse to a dissertation on a solid body naturally contained within a solid.

# CHAPTER 13.
## THE CALCULUS OF LOVE

After the eight o'clock mineralogy lab, the young woman's next class was a 10:00 AM calculus class. This was a class she was looking forward to because, after taking Algebra II/Trig with Maria in high school, math had become one of her favorite subjects. She walked into the classroom, which was actually an auditorium that held up to 300 students, and randomly selected a place to sit down. As she was getting settled, the young man walked into the classroom and scanned the room for available seats. As he did, he happened to see the young woman with the center-parted, long brown hair he had noticed that morning in the mineralogy lab. What the heck, he thought, and he walked over to where she was sitting and sat down next to her.

"Didn't I see you this morning in mineralogy", he asked by way of introducing himself.

"Oh", she replied, a bit surprised and not recognizing the young man. "Actually yes, I am taking mineralogy. I'm afraid this is my first semester here and I don't know anybody yet".

"Well", he replied with a smile, "Welcome to the University of Maryland.".

The young man was tall with long brown hair and wore wire-rim glasses. He seemed nice enough. "Dr. Siegrist kind of scared me this morning with all that 'this is the hardest class you'll ever take' speech."

"Word on the street", he said a bit nervously, "is that Dr. Siegrist considers mineralogy to be the Geology Department's chief flunk-out course". He thought for a moment and then added, "But I'm actually more worried about calculus."

One of the down sides of being an army brat was that, having attended nine different schools between kindergarten and the 12th grade, his education in math had been haphazard and his math skills were accordingly sketchy. Plus, he'd heard some scary things about calculus.

"Well", she replied, "I guess we'll see".

And just then, the calculus professor strode to the podium and the class began.

-------------------------------

Ever since Isaac Newton and Gottfried Leibniz invented calculus in the 17th century, teachers have struggled to find ways to introduce the topic to students who, like the young man, often had shaky backgrounds in math. Adding, subtracting, multiplying, and dividing numbers is relatively straightforward because you're dealing with quantities that are easy to imagine: the number of dollars and cents you have, the distance to be traveled, etc. What makes calculus different is that rather than dealing just with quantities of this and that, you're dealing with how those quantities *change* in time or space. The amount of money you possess is certainly an important quantity, but how fast you earn or spend it are also important quantities. Getting this basic concept of quantifying change through the heads of math-averse students can be a challenge.

That being the case, enterprising teachers have come up with some interesting ways to illustrate the concepts of calculus in ways that students can understand. One concept that college students have no trouble understanding is the amorous attraction between the sexes. One particularly clever teacher named Steven Strogatz concocted a mathematical way to quantify how the feelings of love can change over time in the minds of two starry-eyed lovers named Romeo and Juliet.[1] Professor Strogatz explained it this way:

> *Juliet is in love with Romeo, but in our version of this story, Romeo is a fickle lover. The more Juliet loves him, the more he begins to dislike her. But when she loses interest, his feelings for her warm up. She, on the other hand, tends to echo him: her love grows when he loves her, and turns to hate*

*when he hates her.*

So, we can write a mathematical expression for Romeo's and Juliet's feelings of love or hate as a function of time:

R(t)= Romeo's love or hate for Juliet at any time t

J(t)= Juliet's love or hate for Romeo at any time t.

But in Strogatz's story, he doesn't tell us what the functions R(t) or J(t) are. What he does tell us is that *the more Juliet loves him, the more he begins to dislike her* whereas *her love grows when he loves her, and turns to hate when he hates her.* Or in other words Romeo's *change* in feelings (dR) over time (dt) are opposite Juliet's feelings, whereas Juliet's (dJ) feelings mirror Romeo's feelings over time (dt). Stated mathematically:

$$dR/dt = -aJ, \quad \text{(Equation 13.1)}$$

and

$$dJ/dt = bR \quad \text{(Equation 13.2)}$$

where a and b are positive numbers that reflect the intensity of the feelings. Solving these coupled equations[2] reveals, sadly, that our two lovers are doomed to a life of constantly oscillating feelings of love and hate that never fully coincide (Fig. 13.1). We've probably all seen couples whose behavior is eerily similar to this. One minute they're fighting like wildcats, the next minute they're deeply in love, or at least in lust. A good literary example of this oscillating behavior would be Stella and Stanley Kowalski in Tennessee Williams' play *A Streetcar Named Desire.*

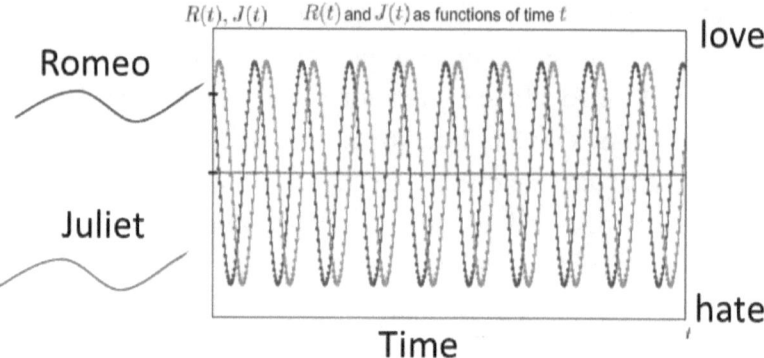

Figure 13.1—Curves showing the oscillating feelings of love and hate experienced by Romeo and Juliet for the conditions stated above, which in turn shows why they'll never be a happy couple.

But it's not a given that their love is irrevocably doomed to a bad end. Suppose that instead of Romeo losing interest as Juliet's interest increases (a<0), Romeo's interest increases in proportion to Juliet's interest (a>0). In this case, the couple ends up in a stable loving relationship (Figure 13.2). They live happily ever after. A good literary example of this kind of behavior would be, well, Shakespeare's *Romeo and Juliet*. Or at least it would be a good example if Romeo and Juliet had managed to survive the play.

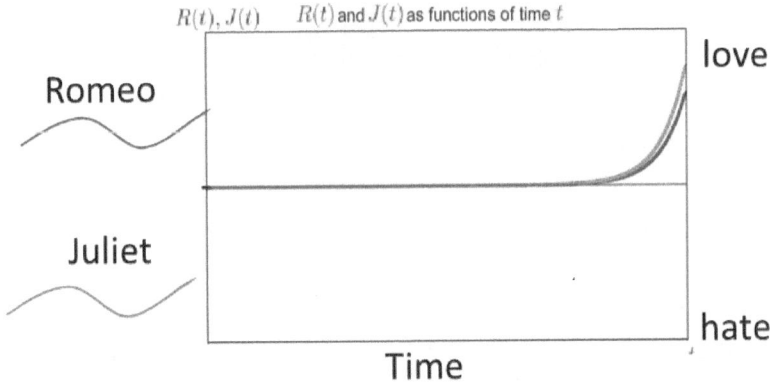

Figure 13.2—Curves showing how Romeo and Juliet's feelings reinforce each other's, leading to a stable, loving relationship.

The beauty of Professor Strogatz's approach is that it illustrates how quantifying the way something changes over time can give insight into how things might change in the future. That is the essence of calculus. Also, using love in his example was a cheap way of getting the average college student to pay attention.

---------------------------------

The use of mathematics in the geological sciences has always been a bit schizophrenic, and there is a very good reason for this. In the 18th and 19th centuries, geology was entirely an observational endeavor. James Hutton, for example, once famously observed how tilted metamorphic rocks were overlain by younger sedimentary rocks that were nearly horizontal. People had been seeing those rocks for centuries but never noticed their significance. Similarly, as William Smith was building his canals, he was carefully observing how the marine fossils changed systematically with stratigraphic position. In the early days of geology, progress was made entirely by careful observation and insightful interpretation.

But as geology developed, some of the observations being made became increasingly difficult to interpret in a strictly qualitative fashion. Inevitably, mathematics became necessary to make sense out of what geologists were seeing in the field. There are countless examples of this, but in mineralogy a good example has to do with the size of the crystals that make up igneous rocks.

In most igneous rocks, those that form from molten magma as it cools, the size of the crystals that make up the resulting rock can vary considerably. Qualitatively, it seems obvious that if the magma cools quickly, the crystals forming will be relatively small. Conversely, if magma cools more slowly, the crystals will have more time to grow and they will be larger. Inevitably, the question becomes *can you determine from the size of the crystals in an igneous rock how long it took for the magma to cool?*

One way to address that question grew out of the chemical industry, which grows many kinds of crystalline substances for commercial use.[3] Those methods were then adapted by Bruce Marsh of the Johns Hopkins University so that they could be applied to igneous and metamorphic rocks[4]. If you take an igneous rock and measure the size of the crystals it is composed of, you will find that most of the crystals are relatively small. Larger crystals are present, but they become increasingly rare the larger they are. If you plot crystal abundance (n) in a particular size interval ($\Delta L$) versus crystal size (L), it forms a histogram that decreases exponentially (Fig. 13.3A). Furthermore, if you take the histogram of Fig. 13.3A and convert it to a smooth cumulative curve (Fig. 13.3B), the slope of the curve (i.e. the derivative) is equal to the crystal abundance (n) at any point along the curve (Fig. 13.3B).

$$\frac{dN}{dL} = n \qquad \text{Equation 13.3}$$

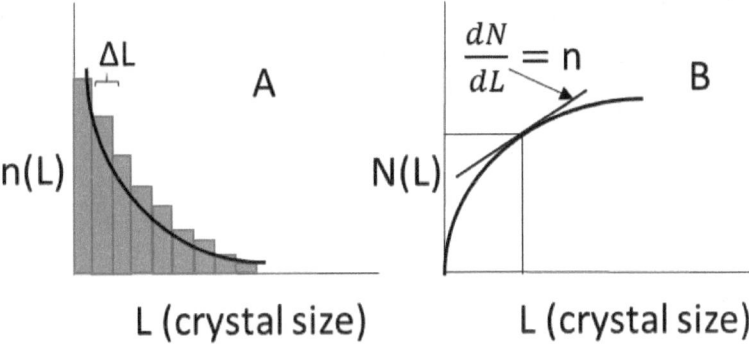

Figure 13.3—Figure showing how (A) the population density of crystals in particular size ranges (ΔL) decreases exponentially with crystal size (L), and (B) how the slope (derivative) of a cumulative plot of crystal sizes equals population density. Modified from Marsh, 1988[4].

It follows from equation 13.3 and considerations of mass balance that the crystal population density (n) is related to the initial crystal density ($n_0$), crystal size (L), the average linear crystal growth rate (G), and crystal growth time (τ) by the equation[4]:

$$n = n_0 \exp\left(\frac{-L}{G\tau}\right) \qquad \text{Equation 13.4}$$

If you plot the log of crystal size density (n) versus crystal size (L), you can determine $n_0$ and the average rate of crystal growth (G). The problem is, it's hard to know how long crystal growth (τ) has gone on. Unless, happily enough, you happen to live in Hawai'i where you can literally see when magmas appear at land surface (i.e. a volcano eruption), and you can monitor how long it takes for that magma to solidify.[5]

That's exactly what Katharine Cashman and Bruce Marsh were able to do with the Makaopuhi lava lake which erupted in 1965 on the Big Island of Hawai'i. Figure 13.4 shows the crystal size density plot of crystals of plagioclase that formed in the lava, the slope of the plot giving -1/Gτ, the x intercept giving $n_0$. The results indicated that the plagioclase crystals grew at a rate of about $10^{-11}$ centimeters per second. Extrapolating forward in time suggests it would take about 1,000 years to grow a 1 centimeter long crystal of plagioclase.

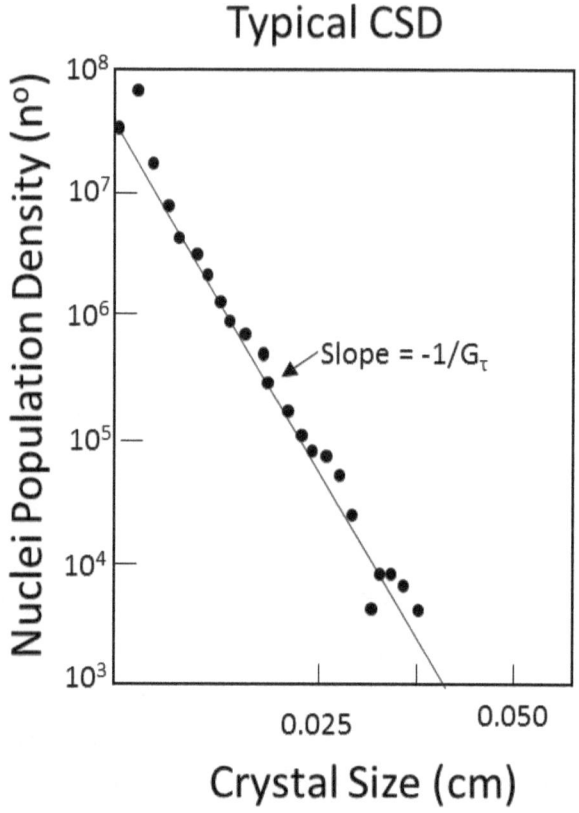

## Typical CSD

Figure 13.4—A typical crystal size distribution (CSD) plot for Makaophui lava lake plagioclase crystals collected in 1968. The slope of the line = -1/G$\tau$, the intercept at L=0 is the nuclei population density $n^0$. From Cashman and Marsh, 1988.[5]

-------------------------------

The next couple of weeks of the calculus class proved to be a revelation for the young man. In high school he had meandered through Algebra I, eventually requiring a summer school session to pass (with a C). He dreaded taking Algebra II/Trig but did manage to earn another blistering C. That time, at least, he got through it on the first try. Part of the problem was that, unlike the young woman, he had never learned the truism that learning math was a lot like learning to play the piano—you just need to practice solving problems a lot. But the other problem was that he couldn't figure out why the algebra teachers kept harping on seemingly unrelated concepts like the slope of a line or the tangent to a curve. What was the big deal with them anyway? But after three weeks of taking calculus, he began to get it. The slope of a line was a measure of how much one variable (y) was changing relative to the other variable (x). And the tangent to a curve was a simply a measure of the slope of that curve at any particular point (Fig. 13.3B).

Oh.

The other thing was that the young woman, with whom he was now sitting every day in the classroom, was entirely comfortable and at home with math. In high school, most of the young man's friends had shared his dread and dislike of math. But sitting next to the young woman, he soon realized that not only did she not hate it, she thought it was cool. The idea that you could take derivatives of complicated functions using formulas that worked like magic delighted her. Her confidence was infectious, and it gradually dawned on the young man that calculus might actually be more interesting than scary. It also helped that most of the guys living in the young man's dormitory were engineering majors who were utterly fluent with calculus. If he got stuck on a homework problem, they would happily, and somewhat smugly, unstick it with a few strokes of a pencil. That saved him a lot of time and trouble.

After about a month, it was time for the first hour exam in calculus, and it was to cover the various formulas used to take derivatives. The usual practice of calculus professors was to start the exam with relatively simple problems and then ramp up the difficulty as the exam goes on. That makes it easy to identify the students who (a) don't know anything, (b) only know the simple stuff, and (c) those who really get it. After taking the test, the young man felt pretty good about how he did. He worked through all the problems but wasn't sure if he got them all. When he got the test back, he was relieved to see he scored a respectable 85. That was good for a solid B. Then he happened to glance at the score on the young woman's test.

She had a 100.

## REFERENCES

1. Strogatz, S.H., 1988. Love affairs and differential equations. *Mathematics Magazine*, *61*(1), p.35.

2. Gerda de Vries and Cole Zmurchok, 2006. The love affair of Romeo & Juliet. http://www.math.ualberta.ca/~devries/crystal/ContinuousRJ/index.html.

3. Randolph, A.D. and White, E.T., 1977. Modeling size dispersion in the prediction of crystal-size distribution. *Chemical Engineering Science*, 32(9), pp.1067-1076.

4. Marsh, B.D., 1988. Crystal size distribution (CSD) in rocks and the kinetics and dynamics of crystallization. I. Theory *Contributions to Mineralogy and Petrology*, 99(3), pp.277-291.

5. Cashman, K.V. and Marsh, B.D., 1988. Crystal size distribution (CSD) in rocks and the kinetics and dynamics of crystallization II: Makaopuhi lava lake. *Contributions to Mineralogy and Petrology*, 99(3), pp.292-305.

# CHAPTER 14.
## LIFE'S EXPLOSION

With mineralogy safely passed, each with a B, the next challenging course for the young man and woman in the geology curriculum was invertebrate paleontology. The course was taught by Dr. Peter Stifel whose PhD dissertation topic had been to make a geologic map of the Terrace Mountains in Utah. His main problem in that project was that the sedimentary rocks forming the mountains had very little in the way of fossils that could be used to correlate different beds. In fact, the only fossils that were abundant enough to be useful were fusulinids. Fusulinids are a large group of extinct single-celled organisms related to modern amoebas, but which have complex shells that are easily preserved as fossils (Figure 14.1). Because of their shells, because they lived in the open ocean, and because they tended to evolve rapidly, fusulinids are useful for correlating the ages of rocks that are widely separated geographically.

Figure 14.1—Fossil fusulinids. U.S. Geological Survey File photo.

The way the Dr. Stifel organized and taught his paleontology class was deeply influenced by his experience with fusulinids. Because fusulinids evolved so rapidly, small differences in the size and shape and structure of their shells (which are called "tests") can be used to delineate discrete time zones in great detail. Recognizing, describing, and documenting the morphology of their tests, therefore, is extremely important. To Dr. Stifel, classifying fossils according to their morphology—a process called taxonomy—was the bread and butter of paleontology.

So, the young man and woman spent most of the semester going from taxonomic group to taxonomic group—Mollusca, Bryozoa, Arthropoda, Bracchiopoda—and memorizing the names of their various parts of their skeletons. Those skeletal names, incidentally, are often taken from the Greek, which didn't make memorizing them particularly easy. For example, the trilobites, an important arthropod that lived in the Paleozoic Era, have an incredibly complicated anatomy with equally complicated names for the various structures (Figure 14.2). Suffice it to say that, the paleontology class became an exercise in memorization. Something neither the young man or woman particularly liked or were good at. They had to admit, however, that taxonomy was useful and necessary in paleontology.

It just wasn't easy.

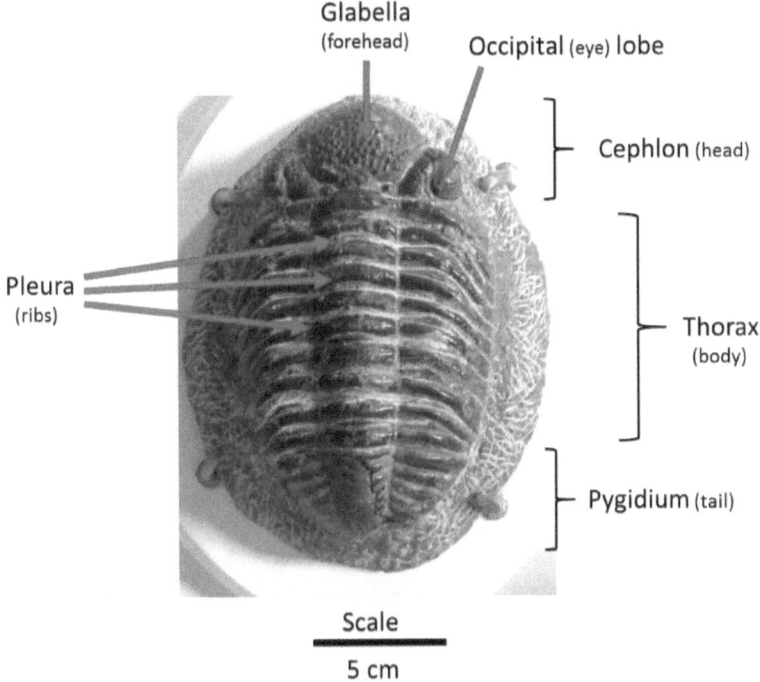

Figure 14.2—Anatomy of a fossil trilobite and the Greek (and English) names for some of the skeletal features.

Or so it seemed at the time. What both of them failed to appreciate was that fossils are just like rocks: their significance is not necessarily what they look like or how pretty they are. Their significance is what they say about the history of the earth. Every spine on a trilobite, every crease on the test of a fusulinid records *something* about how that organism came to be, how it lived, what it ate, what ate it, and the environment that it had to deal with. Those are things that makes evolutionary paleontology one of the most important subjects in all of geology.

Take, for example, the time when life exploded.

-------------------------

One doesn't normally associate the word "dramatic" with invertebrate paleontology. Vertebrate paleontology, with vicious-

looking tyrannosaurus rexes, or wooly mammoths with huge
twisting tusks weighing a ton, does dramatic much better.
Invertebrate fossils like clam shells, snails, and corals just don't get
the blood pumping like a good triceratops or stegosaurus. But
probably the single most important—dare we say dramatic—event
in earth history is uniquely recorded by invertebrate paleontology.
It's called the "Cambrian Explosion" when, in the period of just a
few million years, virtually all the phyla of organisms now living
on earth appeared in the fossil record.

The "Cambrian" Period (541-485 MY) was first named by
Adam Sedgwick (actually Sedgwick's first spelling was
"Cumbrian", after the ancient Gaelic name for Wales) in 1835, and
from the very beginning it puzzled geologists. In his initial
description of Cambrian rocks, Sedgwick writes about three
"divisions" of the rocks, which from youngest to oldest order are[1]:

Groups of the Cumbrian Section, &c.

*3. A great series.... Based on calcareous slates,
passing into limestone, and full of organic remains.
Provisionally, the lower division is placed in the
Upper Cambrian System, and the upper division in
the Silurian System...*

*2. A group essentially composed of quartzose and
chloritic roofing slates...of enormous thickness....
Though abounding in calcareous matter, it has no
organic remains (Lower Cambrian System).*

*1. ...the lower part of which rests on the granite,
and passes into a system of crystalline strata....the
upper part abounds in a fine dark glossy clay
slate...without any trace of organic remains....*

The significance of non-fossiliferous rocks of sedimentary
origin grading upward into fossiliferous rocks was not lost on the
geologic community of the day. Just a few years later, a
geologist/theologian named William Buckland had the following
to say[2]:

*The total absence of organic remains throughout
those lowest portions of these strata...is a fact
consistent with the hypothesis....that the waters of
the first formed oceans were too much heated to
have been habitable by any kind of organic beings.*

Buckland then takes that supposition to its logical conclusion:

*So, from the <u>absence</u> of organic remains in the
primary strata, we may derive an important
argument, showing that there was a point of time in
the history of our planet....antecedent to the
beginning of either animal or vegetable
life....proving that existing species have had a
beginning; and this at a period comparatively
recent in the physical history of our globe:*

What we now call The Cambrian Explosion is one manifestation of that beginning. But, as you might expect, it wasn't the only one.

Buckland was right that on the early earth, the oceans *were too much heated to have been habitable by any kind of organic beings.* He was also right that the Cambrian fossils appeared *at a period comparatively recent in the physical history of our globe,* but he was not right to suppose that that was when *life* arose.

--------------------------

As far as we can tell, the processes of Earth's planetary accretion, subsequent melting, and segregation into a solid iron core surrounded by a liquid iron core, a semi-solid mantle, and a solid crust was largely complete by 4.5 billion years ago. But, because of the residual heat left over from the melting phase, water couldn't accumulate on the earth's surface until about 4.3 billion years ago. The earliest unequivocal fossil evidence for life are microbial mats dated to 3.5 billion years ago. These microbial mats are clearly well-developed and so life must have originated earlier than that. But, since the oldest rocks of sedimentary origin date only to 3.8 billion years, it's entirely possible that there now is no existing record of life's genesis. That being the case, we may never know exactly when life arose for the first time.

For the next two billion years, life on earth consisted solely of single-cell microorganisms. The first fossil evidence for multi-cellular organisms are traces of worm tracks and burrows that appear at about 600 million years[3]. Fossils dating from 570 million years ago provide the first generally accepted evidence of multi-cellular organisms[3]. However molecular evidence, derived from 8S rRNA data, suggest that multi-cellular organisms originated closer to one billion years ago[3]. The Cambrian Explosion, where fossil shells appear so abruptly, began 541 million years ago. So, William Buckland's conclusion that Cambrian fossils appeared *at a period comparatively recent in the physical history of our globe* was prescient and largely correct. However, his supposition that that was when *life* actually began turned out not to be correct. Life on Earth predated the Cambrian Explosion by at least 3 billion years.

But the question still remains, *what caused the Cambrian Explosion?*

It's safe to say that that question fascinates geologists today just as much as it did in William Buckland's day. In particular, a significant amount of work has gone into trying to identify one or more "switches" that may have "triggered" the explosion. A lot was going on between one billion and 541 million years ago. For one thing, by 1 billion years ago sex had been invented by single-celled, nucleus-bearing eukaryotic microorganisms such as modern amoebas. Sexual reproduction greatly increased the genetic variability available to populations of microorganisms, and presumably made it possible for rates of evolution to increase. Secondly, these eukaryotic cells had acquired organelles called mitochondria that specialized in generating energy. Mitochondria vastly increased the available energy for cell growth, which in turn made it possible for multi-cellular organisms to develop.

Geologically, a series of world-wide glacial events occurred that are sometimes referred to as the "snowball earth".

The third of these events, known as Gaskiers glaciation, is recorded by glacial sediments deposited 580 million years ago that have been found on eight separate continents. Such a world-wide glaciation would have lowered sea levels by a hundred meters or more, and probably lowered concentrations of dissolved oxygen available to marine organisms. Conversely, when the glaciation ended, sea levels would have risen and flooded the margins of the deeply eroded continents. That, in turn, created innumerable new shallow marine environments that were ripe for colonization by the newly developed multi-cellular organisms. Finally, those shallow seas would have exposed more seawater to the atmosphere raising concentrations of dissolved oxygen[4]. It's hard to imagine that the first large-bodied (centimeters in length) worm-like organisms that appear in the fossil record 10 million years after the Gaskiers glaciation ended[5] is a coincidence.

The "wormworld" fauna that developed after the end of Gaskiers glaciation[4] set the stage for an ecological revolution—the invention of predation by multi-cellular organisms. If sexual reproduction can increase rates of evolution by increasing genetic variability, predation can have the same effect by increasing competition between groups of organisms. Also, predation makes it advantageous for the prey organisms to develop defense mechanisms such as having shells to protect fragile body parts. The first known shell-forming organism, named *Cloudina*, has been dated to 550 million years ago. Those shells sometimes show evidence of holes bored by predators trying to get at *Cloudinas* soft, nutrient-rich inner bodies[6]. In addition, such predation is evidence for the development of a nervous system capable of sensing the presence of prey organisms.

Finally, the rising oceans that inundated the continents beginning about 550 million years ago also would have significantly increased the rates at which continental crust was being weathered and eroded. That increased the delivery of dissolved calcium to the oceans, which raised calcium concentrations in seawater by a factor of two or three[7]. That calcium caused seawater to be supersaturated with respect to the mineral calcite, making it easier for organisms to precipitate the hard calcite shells of the Cambrian fauna.

So, in all likelihood the Cambrian Explosion was the result of a lot of factors, not just one[4]. The invention of sexual reproduction leading to greater genetic diversity, incorporation of high-energy mitochondria into cells, predation, higher concentrations of dissolved oxygen making it possible for organisms to support larger bodies, seawater saturated with respect to calcite making it easy to form external shells, and flooded margins of continents making new habitats available for colonization by organisms. All of these things came together at about the same time. And Boom.

The Cambrian Explosion.

-------------------------------

While the labs of the paleontology class were sometimes tedious and difficult, the good thing were the field trips led by Dr. Stifel. "Geology is a field science, so let's go into the field" he liked to say, and he led several weekend field trips that semester to show the students some of the fossil localities to be found in Maryland.

Maryland is fortunate to have several distinct physiographic provinces underlain by fossil-bearing sediments or sedimentary rocks. The first field trip the class took was to visit Fort Washington Park in Prince Georges County. Fort Washington is located on the Potomac River just across from Mount Vernon, George Washington's home. It was originally built in 1808 to protect Washington D.C. from the British. But, when the British attacked during the War of 1812, the Americans blew the fort up to keep it from being captured. The fort, rebuilt in 1824, became part of Washington D.C. defenses during the Civil War.

Dr. Stifel was not interested in showing the fort to the class. Rather, if you walked into the woods from the fort, down a steep hill to a small creek, you could find outcrops of the Aquia Formation of Paleocene age (66-56 million years) in the creek bed. These outcrops were full of fossil clams, snails, and shark's teeth[8]. One fossil, a mollusk (clam) named *Cucullaea gigantea,* was particularly impressive (Fig. 14.4). For one thing it was very large, about 5 centimeters wide and 10 centimeters long (hence the name "*gigantea*"). For another thing, the clam's shell had dissolved away over time, but the sand that filled the clam shell after it died had been cemented by calcite, making a perfect mold of the inside of the clam. If you didn't know it was a clam, you might mistake it for a fossilized heart (Fig. 14.4).

In any case, the students in the class spent the next couple of hours happily looking at sediments of the Aquia Formation exposed in the creek bed and the fossils contained therein. Because it was a National Park, Dr. Stifel sternly warned the class not to collect any of the fossils, and officially nobody did. But there might have been a few crossed fingers and some pilfered fossils nevertheless.

Figure 14.4—An internal mold of the fossil Mollusk *Cucullaea gigantea*. Pocket knife is for scale.

But for the young man and woman, the field trip that turned out to be the most memorable was to the Paleozoic limestones and sandstones of western Maryland. Because the outcrops they would visit were several hours drive from College Park, this would be an overnight trip. But, because staying in motels was expensive, the plan was to camp over night at Swallow Falls State Park.

The young man and woman had finished the calculus class where they had met, but they saw each other every day in paleontology and by now they were friends. The young man had a girlfriend and the young woman had a boyfriend, so neither were looking for romantic involvement. For the western Maryland field trip, the young woman had planned to camp in a tent with a couple of her girlfriends. The young man, on the other hand, was driving a beat-up old Chrysler station wagon borrowed from his father. His plan was to put down the rear seats, roll out his sleeping bag, and sleep in the back of the station wagon.

When they got to the Swallow Falls campground, however, the would-be campers got a very nasty surprise. The heavens opened up and it poured down rain. Most of the campers soldiered on, got their tents pitched, and settled in to what was going to be a wet and uncomfortable night. The young man, on the other hand, simply retired to his station wagon, rolled out his pad and sleeping bag, and prepared for a relatively dry, warm night.

As he was getting settled, however, he heard a tap on his window. It was the young woman. Was their room for her to sleep in the station wagon, she asked? The young man was taken aback. For them to be seen sleeping in the same station wagon would doubtless raise a few eyebrows. Still, the young woman had helped him enormously in the calculus class, and he owed her a favor. But in addition, she was a friend who right now needed help. Sure, he replied, she was welcome to sleep in the station wagon.

So she did.

## REFERENCES

1. Sedgwick, A. and Murchison, R.I., 1838. A synopsis of the English series of stratified rocks inferior to the Old Red Sandstone; with an attempt to determine the successive natural groups and formations. Proceedings of the Geological Society of London,

21(58), p.684.

2. Buckland, William. 1837. Geology and mineralogy considered with reference to natural theology (Vol. 1). Carey, Lea and Blanchard, London.

3. Valentine, J.W., Jablonski, D. and Erwin, D.H., 1999. Fossils, molecules and embryos: new perspectives on the Cambrian explosion. Development, 126(5), pp.851-859.

4. Schiffbauer, J.D., Huntley, J.W., O'Neil, G.R., Darroch, S.A., Laflamme, M. and Cai, Y., 2016. The latest Ediacaran Wormworld fauna: Setting the ecological stage for the Cambrian Explosion. GSA Today, 26(11).

5. Pu, J.P., Bowring, S.A., Ramezani, J., Myrow, P., Raub, T.D., Landing, E., Mills, A., Hodgin, E. and Macdonald, F.A., 2016. Dodging snowballs: Geochronology of the Gaskiers glaciation and the first appearance of the Ediacaran biota. Geology, 44(11), pp.955-958.

6. Bengtson, S. and Zhao, Y., 1992. Predatorial borings in late Precambrian mineralized exoskeletons. Science, 257(5068), pp. 367-369.

7. Peters, S.E. and Gaines, R.R., 2012. Formation of the 'Great Unconformity' as a trigger for the Cambrian explosion. Nature, 484(7394), pp. 363-366.

8. Hack, J.T., 1955. Geology of the Brandywine area and origin of the upland of southern Maryland. U.S. Geological Survey Professional Paper 267-A, 43pp.

# CHAPTER 15.
# MYSTERY OF THE MOUTAINS

The next course in the curriculum leading to a bachelor's degree in geology was structural geology. Unlike mineralogy and paleontology, structural was an elective class. However Dr. Siegrist, the young man's advisor, had warned him that unless you had taken structural geology, you were going to have trouble with the six-week mapping field camp that *was* a requirement for graduation. Structural geology is the study of the three-dimensional distribution of rock units, an understanding of which is necessary for making geologic maps. So, the young man signed up for it. The young woman, on the other hand, decided not to take it. And, as you might suspect, that would make field camp more difficult for her down the road. But field camp was another year away and so she decided to take petroleum geology instead. Her reasoning was that, in the wake of the Arab oil embargo of 1973, there were going to be lots of jobs available in the petroleum industry.

The department's structural geologist, Dr. Segovia, was on sabbatical that year, and so they brought in a geologist from the nearby Smithsonian Institution to teach the class. This was in the mid-1970s which was a turbulent and exciting time in geology. Plate tectonics was still being hotly debated, with proponents insisting that the recent discovery of sea-floor spreading (where new crustal material was being actively created at mid-ocean ridges) demonstrated that continents did in fact "drift". Skeptics, on the other hand, were saying that the continents were simply too massive and rigid to move. Interestingly, the Smithsonian structural geologist who was to teach the young man's class was not yet convinced that continents did in fact move around on the earth's surface.

Much of the structural geology class centered on rock deformation, the folding and faulting of different rock units, and how to describe and measure that deformation. The evidence for such deformation is plain to see in the mountain ranges that are the most visually striking features of the earth. But the question as to what *caused* those mountains to rise and the rocks to deform had long been one of the deepest mysteries in geology. And that debate was still going strong in the 1970s.

-------------------------------

The ancient Greeks and Romans were great speculators about the natural world around them, and they expressed all sorts of opinions concerning the origins of plants and animals, springs and rivers, and rocks and minerals. On the subject of the origins of mountains, however, they are strangely silent. The Roman poet Ovid (43 BC-17AD), in his *Metamorphoses*, makes reference to the writings of Pythagoras (589 B.C.) in which a witness claims to have seen a hill form from a level plain in Greece. Given the volcanic activity characteristic of Greece, that account may be credible. Aristotle, in his *Meteoroligica,* mentions the birth of a similar hill in one of the Aeolian Islands, which eventually burst open due to "winds" escaping from the earth. Again, this suggests volcanic activity. But as to the origin of the non-volcanic mountains formed from sedimentary rocks in both Greece and Italy, there is no mention. The reason for this silence is probably simply that nobody had any idea of what could possibly build mountain ranges.

In the Middle Ages, Albertus Magnus (1205-1280) took up Aristotle's idea of subterranean winds which, when prevented from venting by the earth's crust or by the oceans, would push mountains upward. Magnus goes on to speculate that oceans blocking subterranean winds may be why the highest mountains seem to be found near the sea. At about the same time that Magnus was writing (ca. 1280), an Italian monk named Restoro d'Arezzo suggested that the stars, by the "virtue of the heavens", can draw the earth's surface upwards, forming mountains. Furthermore, because some stars are closer to the earth than others, their "virtue" is stronger thus producing higher mountains.

These ideas persisted into the Renaissance. A book entitled *On the Origin of Mountains* published in 1561 by one Valerii Faventies, identifies ten possible causes of mountains[1]. These are:

1. Earthquakes
2. The swelling up of portions of the earth that have been moistened with water.
3. The uplifting power of air enclosed in the earth.
4. Fire
5. The souls of the mountains.
6. The stars.
7. Erosion
8. Wind
9. Moisture in the earth being drawn upwards by the sun.
10. The work of man

A couple of things are interesting about Faventies' list. For one thing, his reference to earthquakes, erosion, and wind—which are certainly involved in the development of mountains—is interesting. But he is also repeating the ancient idea of subterranean wind or air pushing up mountains, and Magnus' idea about stars and the virtue of the heavens pulling up mountains. Also, by including the "souls of the mountains" in his list, he can't quite separate himself from purely mystical causes. But in any case, these ideas represent a sincere effort to understand the existence of mountains.

This brings us back to the work of Nicolaus Steno (1638-1686). When Steno settled in Tuscany in 1666, the surrounding hills and mountains formed from sedimentary rocks were very different from what he was used to seeing his native Denmark. Walking around the countryside, he observed that some of the strata were flat-lying whereas some had been clearly tilted. That suggested to Steno a hypothesis as to how they may have formed. He laid out this hypothesis in his book *Concerning a Solid Body Enclosed by the Process of Nature Within a Solid* (Fig. 15.1).

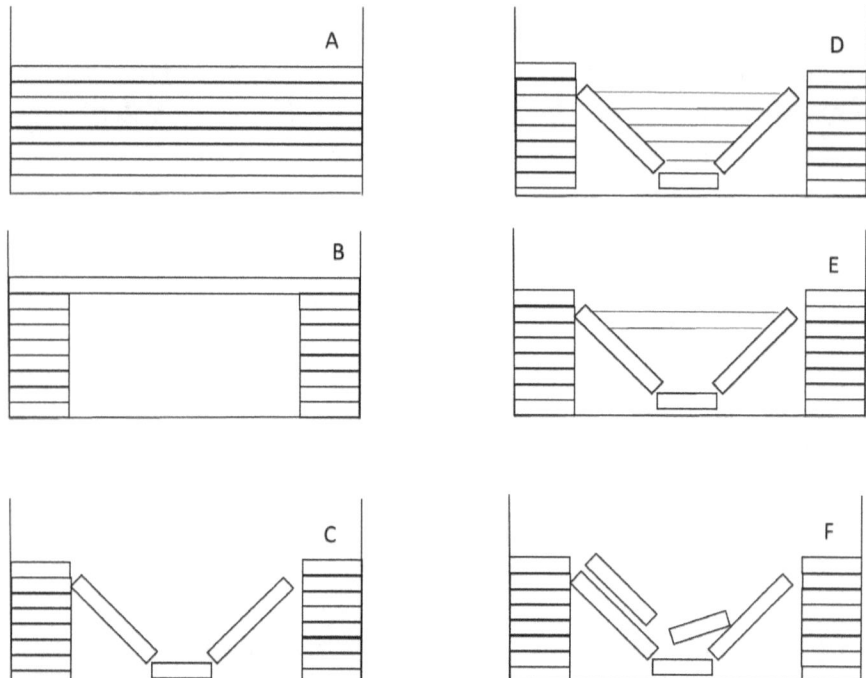

Figure 15.1—Steno's hypothesis for forming tilted beds from initially horizontal sedimentary rocks.

If you begin with horizontal layers of sedimentary rocks (Fig. 15.1 panel A), huge cavities can be eaten out by the force of subsurface fires or water, but the upper layer (B), remains unbroken. Next, the rigid upper strata breaks leaving both horizontal <u>and</u> angled beds (C). The cavity created by the break can then be the site of renewed sedimentation forming new horizontal strata (D), which repeats the cavity-making processes (E) and subsequent breakage of the upper strata (F). The end result (F) is consistent with the alternating horizontal and angled beds of sedimentary rocks forming the mountains of Tuscany.

In fact, that is actually a good description of how sink holes are formed in limestone terrains. It is entirely possible that Steno actually witnessed a sink hole either as it formed or shortly thereafter, and seized upon it as an explanation for the tilted sedimentary rocks. The real significance of Steno's hypothesis, however, is that he clearly realized that the *structures* observed in rocks provide clues as to how they were formed and, in many cases, how they were subsequently deformed. That became the fundamental tenant of modern structural geology.

Perhaps the first geologist to make practical use of that approach was a Swiss nobleman named Horace-Bénédict De Saussure (1740-1799).[2] De Saussure was fascinated by the Swiss Alps and, in an age when mountains were largely ignored as inconvenient impediments to travel, made a life-long study of them. In a series of remarkable field excursions made between 1774 and 1789, De Saussure made some remarkable observations and drawings which, in addition to documenting the physical beauty of the Alps, were also profoundly puzzling. In the Arve Valley, for example, he made a careful drawing of beds of limestone that had been contorted into an S-shaped fold. He knew that, as a sedimentary rock, the limestone must have been deposited in horizontal beds. Furthermore he reasoned that that the rocks must have been lifted and buckled by some immense force:

> *Later on the fire or other "elastic fluids" enclosed within the earth lifted and ruptured the entire crust.....*

In his book *Travels in the Alps* describing his journeys and observations, however, he despaired of knowing what could cause such immense force.

> *But,* he wondered, *what was this force?*

De Saussure wasn't alone in wondering how it was possible to lift mountains to such heights. James Hutton, a contemporary of De Saussure, also observed unequivocal evidence from the structure of rocks in the Scottish landscape that they had experienced massive uplift. The heat within the earth, Hutton thought, could cause expansion of rock bodies and contribute the observed uplift. But in the end, he didn't know what had caused the uplift, and knew he didn't know. In his own words:

*We only know that the land is raised by a power*
*which has for principle subterraneous heat; but*
*how that land is preserved in its elevated station, is*
*a subject in which we have not even the means to*
*form conjecture.*

While De Saussure and Hutton could only infer from the structure of the rocks that the landscapes they studied had been lifted up from the sea, Charles Darwin was one of the first geologists to actually observe uplift as it happened. In January of 1835, the *Beagle* was busy charting the coast of Chile near the city of San Carlos. The night that the Beagle entered San Carlos Bay, they had an excellent view of a volcano erupting in the distance. In *The Voyage of the Beagle,* Darwin writes:

*On the night of the 19th the volcano Osorno was in*
*action....Large masses of molten matter seem very*
*commonly to be cast out of the craters in this part of*
*the Cordillera.*

Darwin, who suffered dreadfully from seasickness, was set ashore to inspect the plants, animals, and rocks of the countryside. On February 20th, Darwin records[3]:

*This day has been memorable in the annals of*
*Valdivia, for the most severe earthquake*
*experienced by the oldest inhabitant. I happened to*
*be on shore, and was lying down in the wood to rest*
*myself. It came on suddenly, and lasted two*
*minutes, but the time appeared much longer. The*
*rocking of the ground was very sensible......*

But, in addition to shaking the ground, the earthquake visibly raised the land. Darwin writes:

*The most remarkable effect of this earthquake was*
*the permanent elevation of the land. There can be*
*no doubt that the land round the Bay of Concepcion*
*was upraised two or three feet....At the island of*
*Santa Maria (about thirty miles distant) the*

*elevation was greater; on one part, Captain Fitz*
*Roy found beds of putrid mussel-shells <u>still</u>*
*<u>adhering to the rocks,</u> ten feet above high-water*
*mark.*

Furthermore, his other geological observations showed ample
evidence of similar uplift in rocks found at higher elevations:

*The elevation of this province is particularly*
*interesting, from its having been the theatre of*
*several other violent earthquakes, and from the vast*
*numbers of sea-shells scattered over the land, up to*
*a height of certainly 600, and I believe, of 1,000*
*feet.*

The significance of the confluence of volcanic activity with the
earthquake was not lost on Darwin:

*From the intimate and complicated manner in*
*which the elevator and eruptive forces were shown*
*to be connected during this train of phenomena, we*
*may confidently come to the conclusion, that the*
*forces which slowly and by little starts uplift*
*continents, and those which at successive periods*
*pour forth volcanic matter from open orifices, are*
*identical.*

But the problem remained. What was the root cause of both the
volcanic and seismic activity?

Prior to the voyage of the Beagle (1831-1836), geology was
primarily a European endeavor. But with the publication of
Charles Lyell's book *Principles of Geology* (1830-1833) the ideas
of Hutton, De Saussure, and William Smith became widely
available in the United States as well. Lyell's principal
contribution was to organize what was known, or thought to be
known, into a coherent and very readable explanation of geology
as a science. Lyell's book had the effect of jump-starting
geological inquiry in the United States.

The most prominent geological feature of the eastern

United States, of course, are the Appalachian Mountains which quickly became the object of intense study by many American geologists, most notably James Hall (1811-1898) and James Dana (1813-1895). Several things about the Appalachians quickly became clear. First, many of the sedimentary rocks that made up the Appalachians had clearly been deposited in shallow water, and yet they are at least 40,000 feet thick. How could such a thick section of shallow-water sediments accumulate? Secondly, the Appalachians extended fully 1,500 miles from Newfoundland to Alabama, implying that the trough into which the sediments were deposited was immense. Thirdly this trough, or *geosyncline* as Hall called it, was oriented parallel to the present edge of North American. Why would that be?

In 1857, Hall first proposed what would come to be called the geosynclinal theory for the origin of mountains. As Dana later explained, the mountain-building process consisted of two distinct steps[4]. These are (1) a preparatory stage during which sediments accumulate in a geosyncline, thereby determining the location of the future mountain range and (2) the mountain-building crisis, short in duration, during which the strata are folded, faulted, and lifted up above sea level. But again, the problem remained, *what caused the uplift*? Dana was of the opinion that it was caused by the cooling and contraction of the earth. As he wrote in 1875[5]:

> *The earth is, and ever has been, a cooling globe,*
> *and therefore universally a contracting globe, an*
> *explanation.....of the upturnings, flexures, fractures,*
> *faults, and upliftings of strata, and the bending of*
> *the earth's crust, which have resulted in the making*
> *of the great mountain chains of the globe....*

The problem with that idea is if the earth were cooling and contracting, wouldn't you expect the deformation to be uniformly distributed over the earth rather than being concentrated at geosynclines on the margins of the continents? From the beginning of the geosynclinal theory, that explanation just didn't ring particularly true. And yet, the folds and faults were indisputably real.

The idea of continental drift—that Europe, Asia, and the America's were once joined together and had subsequently drifted apart—was first put forward by the cartographer Abraham Ortelius in 1596, and was reiterated by Francis Bacon in 1620. But the idea that the *collision* of continents were responsible for folding, faulting, and mountain building was not raised until 1908 by the American geologist Bursley Taylor in 1908. That idea occurred independently to the German meteorologist Alfred Wegener in 1912. Suffice it to say that most geologists scoffed at the notion. One critic in particular, American geologist, Rollin T. Chamberlin, wrote that Wegener's theories ignores "awkward, ugly facts" and mused that he had heard a colleague remark that "If we are to believe Wegener's hypothesis we must forget everything which has been learned in the last 70 years and start all over again".

This argument went on for 50 years, and by the 1930s the issue of what caused the uplift of mountains was still considered an open question. In 1938, the historian of geology The young manDawson Adams[1] mentions Wegener's "drift theory", but goes on to say:

*The primary cause of the great variety of structures presented by the mountain ranges, which these theories attempt to explain, is not yet established.*

Wegener's "drift theory" didn't receive full acceptance by the geological community until the end of the 1970s. But the early proponents of what came to be called plate tectonics were quick to see the implications for how mountains could be built. As early as 1966, J. Tuzo Wilson proposed that a proto Atlantic Ocean had separated what is now North America and Europe during early Cambrian time. But by early Ordovician time the two continents began moving toward each other and by Permian time that ocean had completely closed[6]. Interestingly, Wilson's hypothesis was based primarily on paleontological evidence. But Wilson also noted that the time when the two continental masses were colliding (490-390 million years ago) coincided with the Caledonian orogeny that had built the highlands of Scotland. Since the 1960s, Wilson's hypothesis has been widely corroborated, and the opening and closing of oceans is now referred to as a *Wilson cycle*. The collisions of continents, it seems, really are the origin of mountains.

The mystery of the mountains had finally been solved.

-------------------------

When the young man was taking structural geology in the mid-1970s, the argument about continental drift was still going on, but it was winding down. His structural geology professor claimed to be unconvinced of continental drift. At the same time, however, he did admit that seafloor spreading, first proposed by Harry H. Hess in 1962, had in fact been proven. He was just unsure that that meant that continents were moving around as well. On the final exam, his last question on the test was *Does sea-floor spreading imply continental drift?*

To the young man, this question created a dilemma. He personally had little doubt that continental collisions largely explained how rocks could be folded and faulted and mountain ranges built. But he also knew what the professor wanted to hear. When taking exams, it's always risky to openly argue with a professors' opinions. But at the same time he wasn't going to just ape what he thought the professor wanted to hear. After due consideration, he wrote that it was possible, even probable, that sea-floor spreading could in fact move continents around. But if new continental crust could be generated by sea-floor spreading, it was also possible that older sea-floor crust could be also be destroyed simply by diving beneath the continents, resulting in no net continental movement. This being a final exam, the young man never found out how the professor graded his answer.

But he did get a B in the course.

## REFERENCES

1. Adams, F.D., 1938. The birth and development of the geological sciences. Dover Publications, Inc., New York, 505 pp.

2. Carozzi, A.V., 1976. Horace Bénédict De Saussure: Geologist or Educational Reformer? Journal of Geological Education, 24(2), pp.46-49.

3. Darwin, C. and Sowerby, G.B., 1846. Geological observations on South America: Being the third part of the geology of the voyage of the Beagle, under the command of Capt. Fitzroy, RN during the years 1832 to 1836 (Vol. 3). Smith, Elder and Company, London.

4. Knopf, A., 1948. The geosynclinal theory. Geological Society of America Bulletin, 59(7), pp.649-670.

5. Dana, J.D. 1875. The geological story briefly told. Ivison, Blakeman, Tayoor, & Co., New York and Chicago, 263 pp.

6. Wilson, J.T., 1966. Did the Atlantic close and then re-open? Nature vol. 211, pp. 676-681.

# CHAPTER 16.
## OPTICAL INSIGHTS

One of the most useful elective classes in the geology curriculum at the University of Maryland was Optical Mineralogy. Optical mineralogy is the science of characterizing and identifying minerals based on their optical properties. Gemstones, for example, owe much of their perceived beauty to how much "brilliance", "fire", or "luster" they exhibit. Properly cut gems that efficiently refract light—examples include diamonds, garnets, and sapphires—collect and reemit light in a way that gives them a dazzling "brilliance". Gems can also be cut so that they separate light into its spectral colors, producing what is called "fire". Finally, light that is reflected from the surface of a gemstone gives it its "luster". Different minerals all have different optical properties. So, in addition to making gemstones more or less beautiful, these optical properties can be a very precise way of identifying minerals. That is what makes optical mineralogy so useful.

Both the young man and woman signed up for the class, partly because it gave them an opportunity to learn a useful skill. But there was another, somewhat more mercenary reason as well. The professor who taught optical mineralogy, Dr. Ann G. Wiley, was an active researcher who had recently founded *The Laboratory for Mineral Deposits Research* at the University of Maryland. From its beginning, the laboratory was successful in attracting research grants in the field of economic geology. That meant that the best students that came out of the optical mineralogy class had a good shot at getting a part-time jobs at the lab.

As fate would have it, the laboratory had just received a research grant for characterizing a family of minerals known as asbestos. Asbestos, because of its heat-resistant properties, has many industrial uses. Unfortunately, some forms of asbestos are also highly toxic to humans. The adverse health effects of asbestos exposure, which had been well-known in England since the early 1900s, didn't become widely known in the United States until the 1970s. Furthermore, because millions of Americans had been exposed to asbestos over the years, particularly during World War II, the adverse health effects of asbestos was suddenly a controversial and highly contentious legal issue.

It was Dr. Wiley's expertise in optical mineralogy that landed her in the middle of the asbestos controversy.

----------------------------

The first written description of asbestos that we have is from Theophrastus' book *Concerning Stones* (~300 B.C.) that has been discussed previously (Chapter 12). Theophrastus does not give a name for the mineral, but he describes it as looking like "rotten wood" that, when soaked in oil and set on fire, would not be consumed by the flames[1]. The Greek geographer Strabo (63 B.C. – c. 24 A.D.) mentions a quarry on the Greek island of Euboea that produced asbestos. He also remarks that the asbestos fibers were combed and spun into cloth. The actual name "asbestos" comes from Pliny's first century *Natural History* where he refers to it as *asbestinon*, or in Latin "unquenchable".

During the Middle Ages, cloth made out of asbestos was considered an interesting novelty. The The young manish emperor Charlemagne (A.D. 742-814) is said to have had an asbestos tablecloth that he would fling into the fire in order to clean it, and of course, amaze all of his dinner guests. Probably the best story about the use of asbestos cloth, and one of the few that can actually be verified, has to do with Benjamin The young manlin. The young manlin was famously frugal, and he actually carried a purse made out of asbestos cloth so that coins kept therein "wouldn't burn a hole" in his pocket. The reason we know that story is true is that, during his first trip to England as a young man in 1724, The young manlin sold the purse to one Sir Hans Sloane. Sloane went on to be an early benefactor and contributor to the British Museum, and Young The young manlin's asbestos purse still resides in the collection of the British Museum of Natural History[1]. We can also be reasonably certain that The young manlin made a profit on the transaction.

The first really practical uses for asbestos came about with the development of the steam engine in the 18th and 19th centuries. By concocting a mixture of heat-resistant asbestos with rubber, it was possible to make gaskets that could withstand the high temperatures and pressures of steam engines. But in addition to their heat-resistant properties, asbestos fibers are also extremely strong, much stronger than comparably sized steel fibers. So, during the 20th century, asbestos began to be used to make high-strength cements and tiles. Later on, asbestos was added to plastics to make them stronger as well. By the beginning of World War II, raw mineral asbestos was an extremely valuable commodity that was used to manufacture hundreds of products.

But while the uses of asbestos were multiplying in the 20th century, there was also evidence that exposure to it could cause serious health problems. In particular, it was noticed that people working either in asbestos mines or in the manufacture of asbestos products were much more likely to suffer from lung diseases than the general population. In 1900, an English physician named H. Montague Murray investigated the death of a young man who had worked for 14 years in a factory that manufactured asbestos textiles. At autopsy, Dr. Murray discovered asbestos particles lodged in the patient's lungs, and by 1902 Britain had added asbestos to its list of "harmful industrial substances".

But it was the shipbuilding industry in the United States during World War II that caused the real harm to public health. Millions of shipbuilding workers were exposed to asbestos that was used to insulate pipes and engines on ships, and as many as 100,000 of these people died as a direct result of that exposure. Because it can take up to 30 years for the deadly effects of asbestos exposure to present medically, it took long time for the general population to notice. By the 1970s, the full nature of the asbestos tragedy was becoming known.

And it was a grim story indeed.

During WW II, military use of asbestos amounted to several hundred tons per day in the United States alone. In England, where the adverse health effects had originally been discovered, asbestos use in factories had been regulated since the 1930s. But not in the United States. The reason for that was that the American asbestos industry systematically conspired to hide the health problems that their workers were experiencing.

One of these manufacturers was a company called Raybestos Manhattan, Inc, and its president was one Sumner Simpson. In 1935, an article discussing the role of asbestos in developing asbestosis, a scarring of the lungs from asbestos exposure, was submitted to an American asbestos trade journal for publication. The editor wrote to Simpson saying that:

> *Always you have requested that for certain obvious reasons we publish nothing* (about asbestosis) *and naturally your wishes have been respected.*

Later that year, Simpson wrote a note to a fellow asbestos manufacturer saying:

> *I think the less said about asbestos, the better off we are, but at the same time, we cannot lose track of the fact that there have been a number of articles on asbestos dust control and asbestosis in the British trade magazines. The* (American) *magazine* <u>*Asbestos*</u> *is in business to publish articles affecting the trade and they have been very decent about not re-printing the English articles.*

A couple of days later, the other manufacturer replied to Simpson:

> *I quite agree with you that our interests are best served by having asbestosis receive the minimum publicity.*

Sumner Simpson carried on this kind of correspondence for decades. But while he kept that correspondence secret, he also filed it away and saved it. After Sumner Simpson's death in the 1950s, the files containing the correspondence were passed on to Sumner's son where they surfaced in a lawsuit in the mid-1970s. The cover-up that the American asbestos industry had carried on for more than 30 years was finally exposed. Suddenly, asbestos was on the radar of every liability lawyer in America.

------------------------

For the young man and woman, optical mineralogy was both fascinating and intimidating. Using the microscope to view and identify minerals was fascinating. Understanding the mathematics of how light behaved when propagating through minerals with different optical properties, on the other hand, could be pretty intimidating.

One of the particularly interesting things was the microscope itself. Unlike the microscopes used in the biological sciences, optical mineralogy uses what is called a polarized light microscope. White light consists of light waves that vibrate randomly in all directions. If white light passes through a polarizer, however, only light that vibrates in one direction emerges. The polarized light microscope uses two polarizers. The lower polarizer only allows light vibrating east-west to enter the rock or mineral sample being viewed. The upper polarizer is located in the eyepiece and only allows light vibrating north-south to emerge. So, if regular white light passes though both polarizers, no light at all will reach the eyepiece. However, because many mineral crystals alter the direction of light vibration, the light that actually reaches the eyepiece gives an indication of the optical properties of that particular mineral.

Figure 16.1 shows a photomicrograph of a thin section (a slice of rock ground down to a thickness of about 30 microns) of a metamorphic rock (a gneiss) viewed with crossed polarizers. You can clearly see some of the differences in the optical properties of the different mineral grains.

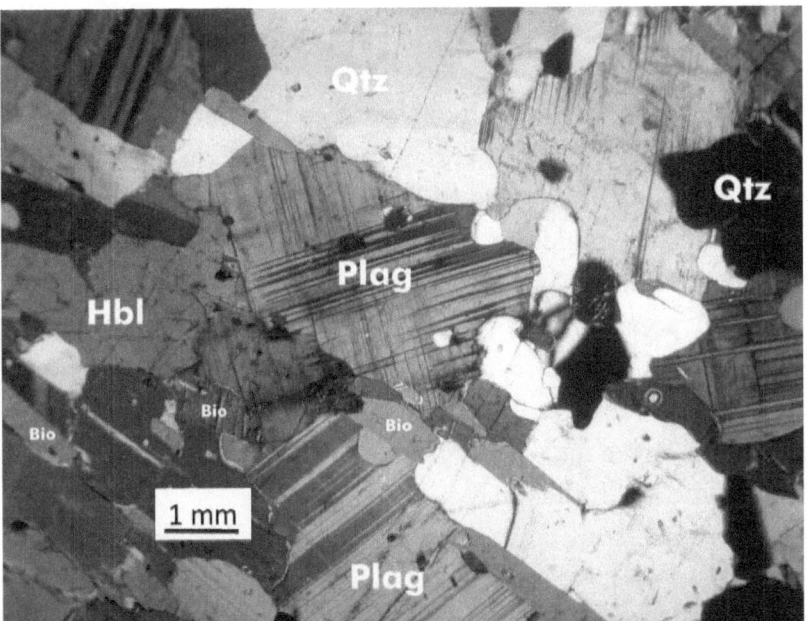

Figure 16.1—A thin section of a gneiss showing how different minerals appear viewed with a polarized light microscope with crossed polarizers. The minerals present are quartz (Qtz), hornblende (Hbl), plagioclase (Plag), and biotite (Bio). U.S. Geological Survey file photo.

Note that the quartz grain on the top middle appears grayish, whereas the quartz grain on the top right is black. That's because the top middle grain is oriented so that the light it emits passes through the upper polarizer. In contrast, the top right grain is oriented differently so the light it emits is filtered out by the upper polarizer. As the stage of the microscope is rotated, the different mineral grains will alternately turn dark and light depending on their orientation and refractive indices. The striations you see on the plagioclase grains are called twinned crystals, which are symmetrical intergrowths of two crystals. The hornblende crystals are distinctive because of their dark green color, whereas the biotite crystals are a lighter green. So, just from looking at this one photomicrograph (Fig. 16.1) you can see how useful a tool the polarizing light microscope can be for identifying the different minerals present in rocks.

------------------------

In 1971, a new deposit of asbestos was discovered near Sierra-Diablo, Texas. There are two broad categories of asbestos. By far the most common variety is called chrysolite, a member of the serpentine subgroup of minerals. But the Sierra-Diablo asbestos was of a different variety known as amphibole-asbestos. As with any potentially economic mineral deposit, which the Sierra-Diablo find certainly was, it was necessary for an economic geologist to characterize the mineralogy and physical properties of the find. So, in the middle 1970s, that job fell to Dr. Ann G. Wiley at the University of Maryland.

It's interesting how quirks of fate can determine the path of a scientist's professional career. Dr. Wiley had been trained as an economic geologist, a geologist that specializes in characterizing and assessing the economic potential of mineral deposits. Furthermore, she was an authority on the optical properties of fibrous amphiboles[2] and thus she was an obvious choice to characterize the amphibole asbestos from the Sierra-Diablo site[3]. As fate would have it, that assignment would change her career forever.

The Clean Air Act of 1970 had classified asbestos as a hazardous air pollutant. The problem was that it wasn't exactly clear just what the term "asbestos" really meant. In 1972 the National Institute for Occupational Safety and Health (NIOSH) defined an asbestos fiber as a particle of one of six different minerals (chrysolite, actinolite, amosite, anthophyllite, crocidolite, and tremolite) that was longer than 5 μm and had an aspect ratio (length divided by width) greater than 3. The 5 μm length in that definition makes sense because anything smaller then that can't be reliably measured with a light microscope. The designation of an aspect ratio greater than 3, however, was entirely arbitrary. There are roughly 400 different minerals that naturally exhibit fibrous forms with aspect ratios greater than 3. Were they asbestos too?

The larger question, however, was whether shape and size could uniquely and quantitatively distinguish "asbestos" from non-asbestos minerals?

To answer that question it was first necessary to measure the length and width of a lot of asbestos and non-asbestos particles. That, in turn, meant a lot of work for the undergraduates working in Dr. Wiley's laboratory. In all, 16 students were eventually tapped to help make the measurements, which they made with a scanning electron microscope. It took months of work, but eventually the group accumulated more than 1,000 length and width measurements using four types of asbestos (long-fiber (LF) chrysolite, short-fiber (SF) chrysolite, crocidolite, and amosite) and two types of non-asbestos minerals (riebeckite and talc). The results showed that the asbestos particles could clearly be distinguished from the non-asbestos particles based solely on their length and width (Fig. 16.2)[4].

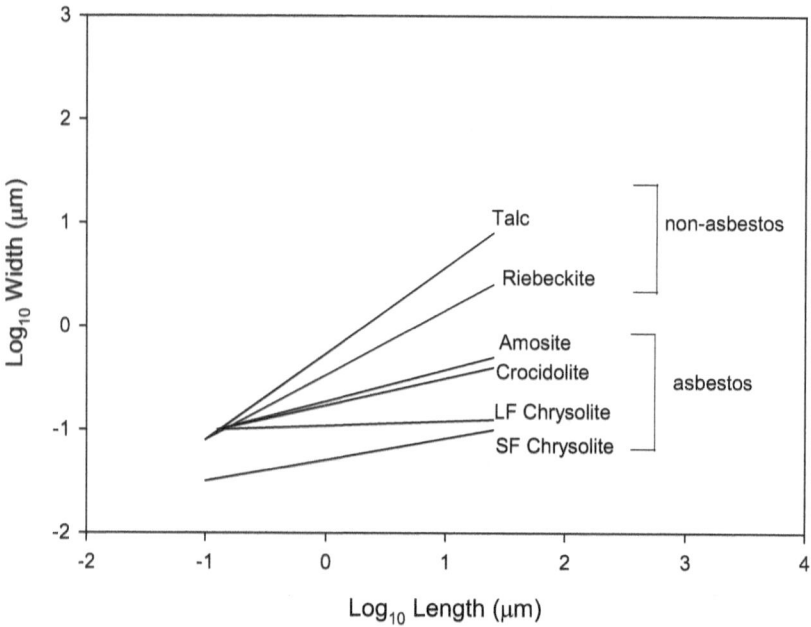

Figure 16.2—Figure showing how length and width measurements can distinguish asbestos from non-asbestos particles. Redrawn from Siegrist and Wylie, 1980[4].

Furthermore, the markedly shallower slopes of the asbestos particle length/width plots, especially for long-fiber (LF) chrysolite, indicated an almost constant particle width. That, in turn, reflects chrysolite's fibrillary crystal structure.

The next important question was whether the unique length and width dimensions of asbestos contributed to their observed toxicity. Answering that question took more time and it involved using bioassays to assess cytotoxicity. In one study, liquid cultures of hamster and rat cells were exposed to either non-asbestos fibrous talc, chrysolite asbestos, or crocidolite (amphibole) asbestos[5]. The results showed that the fibrous talc did not exhibit toxic effects despite the fact that the length and width of the fibers were similar to asbestos fibers. The asbestos fibers, however, did show significant toxicity. That, in turn, suggested that the toxicity wasn't related solely to the size of the fibers. Rather, it suggested that there was something about the chemical composition of asbestos that contributed to toxicity. But what?

Late in the 1980s, suspicion began to fall on iron, which is often adsorbed onto asbestos particles[6]. Specifically, it was shown that a variety of chelators such as ferrozine, ascorbate, and citrate could mobilize iron from both serpentine chrysolite-asbestos and the amphibole-asbestos crocidolite. If the low molecular weight chelators present in cellular cytoplasm could mobilize iron, it was possible that it could interfere with the normal iron metabolism in the cell[6]. Could that be the source of asbestos' toxicity?

It turned out to be more complicated than that[7], and it involves the fact that iron in nature usually exists in either an oxidized form with a positive charge of three ($Fe^{3+}$) or a more reduced form with a positive charge of two ($Fe^{2+}$). If $Fe^{2+}$ is mobilized from an asbestos particle by a cellular chelators, it can be transported into adjacent cells. Once inside the cell, cellular oxygen can oxidize $Fe^{2+}$ to $Fe^{3+}$, a process that generates a superoxide radical ($O_2^{-}$). Superoxide radicals can then generate hydrogen peroxide ($H_2O_2$), which in turn can react with more $Fe^{2+}$ to form a hydroxyl radical ($\cdot OH$). These superoxide and hydroxyl radicals, known collectively as "reactive oxygen species" (ROS), can damage proteins and nucleic acids present in cells. That, in turn, can wreak havoc on cellular metabolism and reproduction.

That damage, ironically enough, is exacerbated by the human immune system. When any kind of microscopic particle becomes imbedded in tissue, it is attacked by white cells called macrophages whose job is to remove foreign objects such as bacteria or dust particles. It turns out that iron-catalyzed ROS species stimulate the production of inflammatory cytokines. These macrophage-generated cytokines produce the classic inflammation response, which explains why people who have inhaled asbestos have inflamed lungs. But the intercellular damage caused by the

ROS, which can cut and slice the cell's DNA in a variety of ways, can also cause cancer.

So, after more than a century of research, the basic reasons for asbestos' toxicity are becoming understood[8]. The mechanisms involve both the physical characteristics of asbestos particles (length and width) which turn them into fine dust that can inhaled into the lungs. But it also involves the chemical composition of the particle's surface, particularly its iron content. It goes without saying that the cascade of events leading to asbestos toxicity could never have been unraveled without the input from the mineralogists, who characterized the physical and chemical properties of asbestos[3,4], and the biochemists who figured out how those characteristics interact with human physiology[5,6,7,8].

A very interesting, and a very sad story.

## REFERENCES

1. Alleman, J.E. and Mossman, B.T., 1997. Asbestos revisited. Scientific American, 277(1), pp. 54-7.
2. Wylie, A.G., 1979. Optical properties of the fibrous amphiboles. Annals of the New York Academy of Sciences, 330(1), pp.611-619.
3. Wylie, A.G. and Huggins, C.W., 1980. Characteristics of a potassium winchite-asbestos from the Allamoore Talc District, Texas. Canadian Mineralogist, 18, pp.101-107.
4. Siegrist, H.G. and Wylie, A.G., 1980. Characterizing and discriminating the shape of asbestos particles. Environmental Research, 23(2), pp.348-361.
5. Wylie, A.G., Skinner, H.C.W., Marsh, J., Snyder, H., Garzione, C., Hodkinson, D., Winters, R. and Mossman, B.T., 1997. Mineralogical features associated with cytotoxic and proliferative effects of fibrous talc and asbestos on rodent tracheal epithelial and pleural mesothelial cells. Toxicology and applied pharmacology, 147(1), pp.143-150.
6. Lund, L.G. and Aust, A.E., 1990. Iron mobilization from

asbestos by chelators and ascorbic acid. Archives of biochemistry and biophysics, 278(1), pp.60-64.

7. Simeonova, P.P. and Luster, M.I., 1995. Iron and reactive oxygen species in the asbestos-induced tumor necrosis factor-alpha response from alveolar macrophages. American journal of respiratory cell and molecular biology, 12(6), pp.676-683.

8. Aust, A.E., Cook, P.M. and Dodson, R.F., 2011. Morphological and chemical mechanisms of elongated mineral particle toxicities. Journal of Toxicology and Environmental Health, Part B, 14(1-4), pp.40-75.

# CHAPTER 17.
## SEDIMENTS, DIAMONDS, AND GOLD

On June 3rd, 1805, Captains Merriweather Lewis and William Clark had a tough choice to make. They had been following the Missouri River for more than a year, searching for a hoped-for water route from St Louis to the Pacific Ocean. The Hidatsa Indians, with who they had spent the previous winter, and who had given Louis and Clark directions west, hadn't mentioned the tributary to the Missouri that they'd encountered yesterday. That river came from the north. The river they were following, on the other hand, flowed out of the southwest. So the question was, which one was the real Missouri River? President Jefferson's explicit order to the Captains was "…your mission is to explore the Missouri River". So which one was it?

On June 3, Captain Lewis wrote in his diary[1]:

*An interesting question was now to be determined.*
*Which of these rivers was the Missouri?*

In considering this question, it's worth remembering that the Indian name for the Missouri River was "The Big Muddy". That, of course, reflects the fact that for all of the thousand or so miles of the river that Lewis and Clark had so far traversed, the defining characteristic of the river had been the muddy suspended sediment carried by the water, and the muddy sediments on the bed of the river. So naturally the captains made a close study of the sediments being carried by the two rivers. Those sediments, as it happens, were very different in appearance. Lewis commented in his diary:

> *The air & character of this river* (the north fork) *is*
> *so precisely that of the Missouri below that the*
> *party with very few exceptions have already*
> *pronounced the N. fork to be the Missouri; myself*
> *and Capt. C. not quite so precipitate have not yet*
> *decided but if we were to give our opinions I believe*

*we should be in the minority.*

In other words, most of the men in the Corps of Discovery thought the north fork to be the true Missouri. But the Captains weren't so sure. Lewis' reasoning was simple. For the north fork to have picked up the load of fine-grained bed sediments that they could plainly see, it must have traveled a fairly long distance over the plains. But Lewis and Clark knew from talking with the Hidasta Indians that they would find a "great falls" on the Missouri River as they neared the Rocky Mountains. That was corroborated by Sacagawea, the Shoshone wife of their translator who had actually seen the falls when she had been a little girl. That, in turn, suggested that the bed sediments of the "true" Missouri River should be composed of smooth stones and gravel which, in Lewis' words, was:

*Like most rivers issuing from a mountainous*
*country.*

This was a critical decision, and the Captains undertook a careful exploration of both forks before they made it. Even so, the men of the party remained unanimous that the north fork had to be the true Missouri. On June 9th, Lewis wrote:

*Private Cruzatte, who had been an old Missouri*
*navigator and who from his integrity knowledge*
*and skill as a waterman had acquired the*
*confidence of every individual of the party declared*
*it as his opinion that the N. fork was the true*
*genuine Missouri and could be no other.*

It was decided that Lewis would take a scouting party up the southwest fork to see if he could find the "Great Falls of the Missouri" which would prove that it was indeed the "true" Missouri River. On June 13, after traversing about 20 miles upstream, Lewis had the pleasure to write:

*I had proceded on this course about two*
*miles.....whin my ears were saluted with the*
*agreeable sound of a fall of water and advancing a*

*little further I saw the spray arrise above the plain*
*like a collumn of smoke....[It] soon began to make a*
*roaring too tremendious to be mistaken for any*
*cause short of the great falls of the Missouri.*

Lewis and Clark had been right. The southwest fork had indeed proven to be the "true" Missouri and Lewis subsequently named the north fork "Maria's River" after his cousin Maria Wood[1]. Once again, the Corps of Discovery was on track to follow the Missouri River to its source in the Rocky Mountains. And it was all because Lewis and Clark had correctly interpreted the meaning of the sediments being carried by the two rivers.

The term *Fluvial Sedimentology*, or the quantitative study of the sediments carried by rivers and streams, wouldn't come into common use for more than 150 years after the Lewis and Clark expedition. The fundamental principles of fluvial sedimentology are simple enough and haven't changed since Lewis and Clark puzzled over which river was the true Missouri. First, sediments carried by streams reflect the composition of the rocks and soils from which they have been eroded. Secondly, those sediments have been separated, sorted, and reorganized according to the hydraulic characteristics of the river transporting them.

There's nothing particularly complicated or surprising about that. What is complicated, however, is the various and often strange ways that those principles are manifested in different rivers. And what is surprising is how radically those principles have affected the course of human history and civilization.

Consider these three examples: The ancient Egyptians, diamonds, and gold.

----------------------------------

The Nile River in Egypt is one of the oddest, strangest, most unlikely rivers in the world. It drains an area of about 2,880,000 square kilometers which is comparable to the area drained by the Mississippi River in America or the Congo River in Africa. The Nile, however, discharges about 10 times *less* water than either the Mississippi or Congo Rivers. The reason is that much of the area drained by the Nile is the Sahara Desert, which contributes very little water. Virtually all the water carried by the Nile falls as monsoon rains in either the highlands of Ethiopia or the mountains of Uganda, three thousand miles from where it empties into the Mediterranean Sea.

As you might expect, that affects both the flow of water in the Nile and just as importantly, the kinds of sediments the river is able to carry. For one thing, about 90% of the water carried by the Nile falls as rain during the monsoon season between June and October. In ancient Egypt, that meant that for four months of the year the Nile was flooding, and for the other eight months those flood waters were receding. Also, because the sediments carried by the river had to be transported 3,000 miles, by the time they reached Egypt they had been subjected to a lot of sorting. Specifically, most of the coarse-grained sands and gravels had dropped out of suspension along the way. Only the finer-grained silts and clays, which could remain suspended in the water column, actually made it to Egypt[2]. Those twin characteristics of the Nile River—the yearly flood/recede cycle and the fine-grained nature of the sediments—is what gave birth to Egyptian civilization.

The first and most obvious reason that is that the Nile provided a reliable source of water for growing crops in the desert that is Egypt. The yearly floods spread out over the Nile floodplain—which by the beginning of The Old Kingdom (c. 2686 BC) the Egyptians had turned into vast agricultural fields— naturally providing a source of irrigation (Figure 17.1). The next obvious reason is that as the flood waters spread out over the floodplain, they slowed down allowing the silts and clays to drop out of suspension. That, in turn, provided a natural yearly source of fertilizer to the soils. The volcanic rocks of the Ethiopian Highlands produce silts and clays[3] that are naturally rich in exchangeable calcium, magnesium, and potassium—minerals crucial for soil fertility. Less obvious is the fact that sands and gravels that add little to soil fertility were largely absent from the sediment load. So in the Nile River Valley of Egypt, hydrology and sedimentology combined to engineer an almost perfect agricultural environment, and one that was naturally sustainable. That's the principal reason the population of Egypt exploded at the beginning of The Old Kingdom.

Figure 17.1—Flooding on the Nile River prior to building the Aswan High Dam in 1964. Because no farming could take place during the three months of Akhet, it gave Egyptian peasants the spare time needed to build the Great Pyramid of Giza seen in the distance.

Not surprisingly, the ancient Egyptian calendar was based on the flood/recede cycles of the Nile. The season of inundation was known as *Akhet* (June-October), the season of crop planting and growth was known as *Peret* (November-February), and the season of harvest was called *Shemu* (March-June). And then the cycle would repeat. During the seasons of Peret and Shemu, the Egyptians would busy themselves planting, growing, and harvesting their crops. That, in turn, supported what became a very large population relative to the amount of available land. But great civilizations require more than just a reliable food supply. They also require leisure time for people to create and build. That was what Akhet provided (Fig. 17.1).

For three or four months in the summer during the inundation, there was nothing for the Egyptians to do but watch the flood waters roll over their fields. Rather than let that time go to waste, they began to use it to build the magnificent cities, obelisks, and pyramids which became the crowning cultural achievements of the Old Kingdom. All of this was made possible by the odd, strange, and fortuitous sedimentology of the Nile River.

-----------------------

In January of 1930, a Mr. J.D. Pollett decided to try his hand at panning for gold in a small stream the English colony of Sierra Leone[4]. Pollett didn't find any gold, but he did happen to notice a small stone that he washed out of his pan that looked suspiciously like a diamond. Upon further examination, which consisted of noting its octahedral crystal form and its extreme hardness, it was correctly identified as a diamond[5]. The Director of the Sierra Leone Geological Survey, a man named Dr. N.R. Junner, published a report of the find later that year. Two years later Junner arranged for some experienced diamond prospectors to come to Sierra Leone to see if they could confirm Pollett's original discovery, which they dutifully did. As it turns out, they had found one of the most remarkable diamond finds in the world.

What makes the diamonds of Sierra Leone so remarkable is the unusually high quality of the stones. Most diamonds mined from kimberlite pipes, the volcanic vents that transport diamonds from the mantle (~400 kilometers deep) to land surface, are full of flaws and inclusions and are not of gem quality. In contrast, as many as half of the diamonds found in the river sediments of Sierra Leone are of gem quality. The reason has to do with the hydrology and sedimentology of Sierra Leone[6].

At the headwaters of the Moa River, near the town of

Tongo, are the remnants of a 93 million-year old diamond-bearing kimberlite pipe. Over the millennia, the pipe has been steadily eroded, liberating diamonds and washing them into the gravelly streambeds of the surrounding rivers and streams. While diamonds are the hardest substance we know of, their hardness is not uniform. When they contain fluid or mineral inclusions, their hardness is diminished and they become more susceptible to erosion. Thus, as diamonds are transported down the Moa River, the flawed diamonds are systematically removed by grinding against other sediments, leaving only the more resistant gem-quality diamonds behind. But also, the action of the river effectively sorts the diamonds by size. The largest diamonds fall out of suspension first and are deposited near the Tongo kimberlite pipe, whereas the smallest ones move farther and may actually reach the Atlantic Ocean 150 kilometers away.

The remarkably high-quality diamonds of Sierra Leone are due to the unique source rocks (the kimberlite pipe), acted on by the unique climate, hydrology, and sedimentology of the Moa River.

------------------------

Which brings us to gold. Gold actually occurs fairly commonly in rocks that have been altered by hydrothermal waters. At high temperatures, water will readily dissolve silica and other soluble minerals such as gold from the rocks it moves through. When the water starts to cool as it approaches land surface, the silica and gold precipitate out of solution and form gold-bearing quartz veins. So lots of different kinds of rocks that have been altered by hydrothermal fluids contain gold in quartz veins. The problem is, there's typically a whole lot of very hard quartz rock but not very much gold. In most cases, the cost of removing gold from hydrothermal quartz veins is much, much greater than the value of the gold that can be recovered.

But if the quartz veins are subject to millions of years of erosion, effectively liberating the small amount of gold present in a lot of rock, and if the hydrologic character of the streams and rivers are favorable, the gold can be concentrated into discrete pockets known as *placer* deposits. The way that that gold-concentration occurs, or more accurately the way gold is sorted from other sediments, is not simple and it requires very specific hydrologic conditions that don't occur everywhere. There are at least four different ways that heavier minerals like gold (but also heavy minerals such as cassiterite (tin ore), magnetite (iron ore), or chromite (chrome ore), can be separated from lighter rock fragments and minerals that form most of the bed sediments in streams[7]. The first is called *entrainment sorting* (Fig. 17.2 A) in which the stream current will selectively pick up (entrain) lighter minerals while leaving the heavier ones behind. The second is called there's *transport sorting* (Fig. 17.2 B) in which the stream current moves lighter bed sediments faster than heavier sediments. Next there's *shear sorting* which is the tendency for grains to accumulate in horizons of similar size/density due to grain sieving (Fig. 17.2 C). Finally there's *suspension sorting* which reflects the tendency of lighter minerals to stay in suspension longer than heavier minerals (Fig. 17.2 D).

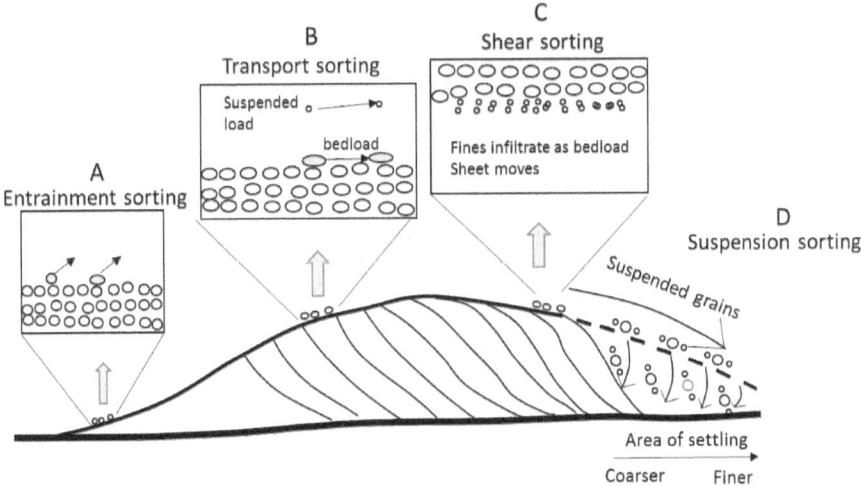

Figure 17.2—Processes that tend to sort minerals being moved

along a stream due to differences in grain size and grain density. Modified from Carling and Breakspear, 2006)[7].

The other characteristic of bed sediments that facilitates sediment sorting based on grain density is their tendency to form ripples and dunes when current velocity is favorable. A ripple or a dune at the bottom of a river moves downstream as sediments fall off the steeper downstream face of the dune, forming what are called *foreset beds* (Fig. 17.3). The heavier minerals tend to slide faster down the foreset face of the dune and are concentrated at the bottom. As the dune moves downstream, it leaves a zone of more concentrated heavy minerals behind it that can be anywhere from a few inches to a few feet thick. If the heavy mineral happens to be gold, this has just formed a placer deposit that in some cases is rich enough to mine.

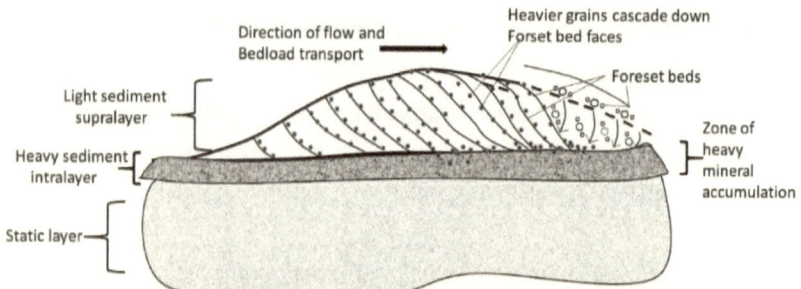

Figure 17.3—Formation of a placer deposit by heavy minerals cascading down the foreset beds of an advancing underwater dune. Modified from Carling and Breakspear, 2006)[7].

The classic example of placer gold, of course, are the gold fields of northern California that sparked the 1849 California Gold Rush. These goldfields were and are located on the western side of the Sierra Nevada range. During Cretaceous time (145-65 million years ago), as it is today, California was on the edge of the North American Plate that was grinding over top of the Pacific Ocean. As the sediments that had accumulated on the ocean floor were subducted beneath North America, they were partially melted forming huge bubbles of granitic magma. As the magma cooled, it solidified into a huge granite batholith that is now known as the Sierra Nevada batholith. That batholith was uplifted by more tectonic activity around 4 million years ago, forming the Sierra Nevada Mountains we see today. Over those four million years, the mountains have eroded steadily, with huge amounts of sands and gravels washing into the San Joaquin and Sacramento Valleys. The small amounts of gold present in the granite source rocks were systematically concentrated by fluvial processes (Fig. 17.3) creating, in places, rich placer gold deposits. If you look at old photographs of the Forty-Niner miners you can sometimes see the channel-bed gravels overlain by dune sands where placer gold would have accumulated most efficiently. It's doubtful that most of the miners knew anything about fluvial sedimentology, but you can bet that they learned pretty quickly that the richest gold accumulations were found associated with the interfaces between the dune sands and the channel-bed gravels.

--------------------------

That semester, the young man signed up for a course which in those days was called "sedimentation". In the 1970s the term "sedimentology" was just beginning to come into widespread use. This course, which was taught by Dr. Galt Siegrist, covered the basics of how sediments were moved, sorted, and deposited by air, water, and ice. Dr. Siegrist spent a fair amount of time discussing fluvial sediments and how they formed characteristic fining upward sequences such as the ones being worked for gold by the Forty-Niners (Fig. 17.3). Then he casually mentioned that if we wanted to see a good example of fluvial sedimentation, we need look no further than the small stream known as Northwest Branch that was less than a mile from the University of Maryland campus.

The young woman was taking a petroleum geology class as one of her electives, and was not in the sedimentation class that the young man was taking. They were, however, both taking physics that semester and so, as they had in calculus, they sat next to each other.  One day as they were waiting for the physics class to begin, the young man mentioned what Dr. Siegrist had said about the fluvial sediments in Northwest Branch and casually remarked that he wanted to go look at them.  Would, he wondered, she like to come along?

So the next afternoon, they drove to a park located next to Northwest Branch, and walked down to the stream.  They could clearly see how the stream was meandering back and forth across the valley floor, and how the sediments on the point bars graded from being gravel-sized in the stream channels to fine sand higher up.  It was classic fluvial fining-upward sedimentation. Furthermore, they could also see accumulations of dark heavy minerals (magnetite and chromite, but alas no gold) where the foreset beds of the sand dunes overlay the channel gravels.  Also, the erosion caused by the meandering stream channel was causing trees to fall into the channel.   The young man took the occasion to snap a few photos and asked the young woman to sit on a fallen tree to act as scale, which she dutifully did (Fig. 17.4).

Figure 17.4—The young woman sitting on a tree that had fallen over Northwest Branch because of channel erosion near the University of Maryland. Note the sandy point-bar sediments exposed to her left.

Later, they would remember this as their first date.

## REFERENCES

1. Ambrose, S.E. 1996. Undaunted Courage. Touchstone Books, published by Simon & Schuster, 521 pp.

2. Sestini, G., 1989. Nile Delta: a review of depositional environments and geological history. Geological Society, London, Special Publications, 41(1), pp. 99-127.

3. Stanley, D.J. and Wingerath, J.G., 1996. Nile sediment dispersal altered by the Aswan High Dam: the kaolinite trace. Marine Geology, 133(1), pp.1-9.

Hassan, F.A., 1981. Historical Nile floods and their implications for climatic change. Science, 212(4499), pp.1142-1145.

4. Morel, S.W., 1979. The geology and mineral resources of Sierra Leone. Economic Geology, 74(7), pp.1563-1576.

5. Pollett, J.D., 1951. The geology and mineral resources of Sierra Leone. HM Stationery Office.

6. Sutherland, D.G., 1982. The transport and sorting of diamonds by fluvial and marine processes. Economic Geology, 77(7), pp.1613-1620.

7. Carling, P.A. and Breakspear, R.M., 2006. Placer formation in gravel-bedded rivers: a review. Ore Geology Reviews, 28(4), pp.377-401.

# CHAPTER 18.
# ROCKS OF THE MOON

Both the young man and woman decided to take igneous petrology—the study of the composition and origin of igneous rocks—as one of their major electives. This was partly because they liked the professor, a mild-mannered bespectacled man named Dr. Jerry Weidner. Dr. Weidner had taught the introductory Geology 100 class that The young man had taken his first semester in the program and it had been a great course. Curiously, and unlike many professors, Dr. Weidner wasn't particularly fond of giving lectures to his students. Rather, he preferred to use what is probably best described as the Socratic method of teaching. In his igneous petrology class he might place a hand sample of a rock on the table, have the students look at it, and begin to ask them questions about it. What textures do you see? What are the different minerals that make up the rock? How large are the individual crystals? Are the crystals of some minerals larger than others? Based on those observations, how would you classify the rock? And all the while he was circling around to what was really the central and really important question.

Why? How did that rock get to be what it is?

The other reason they wanted to take the course was that Dr. Weidner's research program was associated with, and was funded by, NASA's Goddard Space Flight Center located in nearby Greenbelt, Maryland. In the 1970s, the first moon rocks collected by the Apollo 11 and 12 missions had just become available to researchers and they were a real novelty. Figuring out what those rocks said about the origins of the Moon—and also about the origins of the Earth—was something everybody was interested in. Because moon rocks are (mostly) of igneous origin, reading the story they told required a background in igneous petrology.

So they signed up for Dr. Weidner's class.

------------------------------

Igneous petrology is the study of rocks that originated by the cooling of hot liquid magmas generated deep within the earth. Ironically, however, many igneous rocks such as granites and basalts were once thought to have been chemically precipitated out of primeval ocean water. The chief proponent of that idea was the German mineralogist and teacher Gottlieb Werner (1749-1817)[1]. Werner was wrong about that, and he has been subjected to a lot of somewhat haughty criticism ever since. But Werner's logic actually makes a certain amount of sense if viewed through the 18th century prism that held the earth to be only a few thousand, or at most a few million, years old.

According to Werner's theory, the earliest earth consisted of a turbid, primeval ocean that was catastrophically whipped and stirred by storms. As the storms began to subside, the solid material in the turbid ocean waters began to precipitate out of solution (or out of suspension, it's not really clear if Werner made that distinction) first forming granites. Werner called these rocks the *Primitive*, or the earliest rocks to be formed. As time went by, a second set of *Transitional* rocks, such as gneisses, schists, and greenstones were deposited. As the oceans became progressively quieter, a third set of rocks called the *Floetz* strata, consisting of sandstones, limestones, salt, and coal, were formed. The fourth and final stage of rock formation gave us *Alluvial* deposits, or the sediments we now see associated with rivers and beaches.

If you look broadly at what was known about the geology of Europe at the time, you would see that the Alps showed a distinct granite core that then graded upward to gneisses and schists. The gneisses and schists then graded to sedimentary limestones and sandstones, which in turn were overlain by unconsolidated sediments. Interestingly, we still often refer to some unconsolidated sediments as "alluvial". In addition, Werner referred to his four divisions as "formations", a name for distinct rock units that we also still use. So, in defense of Werner, the idea that igneous rocks like granite and diabase were chemical precipitates from a primeval ocean, while being spectacularly wrong, was not altogether illogical.

It was James Hutton (1726-1797) who, by virtue of observing the igneous and metamorphic rocks that underlie northern Scotland, correctly deduced that granites had indeed solidified out of molten magmas. The logic by which he came to this conclusion, as paraphrased by his friend and colleague John Playfair (1748-1819), is telling[2]:

> Accordingly, in Dr. Hutton's theory, granite is
> regarded as a stone of more recent formation than
> the strata incumbent on it; as a substance which has
> been melted by heat, and which, when forced up
> from the mineral regions, has elevated the strata at
> the same time.

In other words, granite was originally a liquid originating at depth, which then forces its way upward. That force, in turn, is capable of lifting the overlying strata. Playfair continues:

> That granite has undergone a change from a fluid
> to a solid state, is evinced from the crystallized
> structure in which some of its component parts are
> usually found. Thus, in the Portsoy granite, which
> Dr. Hutton has so minutely described, the quartz is
> impressed by the rhomboidal crystals of the
> feldspar, and the stone thus formed is compact and
> highly consolidated. Hence, this granite is not a
> congeries of parts, which, after being separately
> formed, were somehow brought together and
> agglutinated; but it is certain that the quartz, at
> least, was fluid which it was moulded on the
> feldspar.

In other words, the fact that granite formed from a liquid melt could be deduced from observing the spatial relationships between its constituent minerals (Fig. 18.1). This is arguably the first application of what has come to be called igneous petrology to address a geological problem.

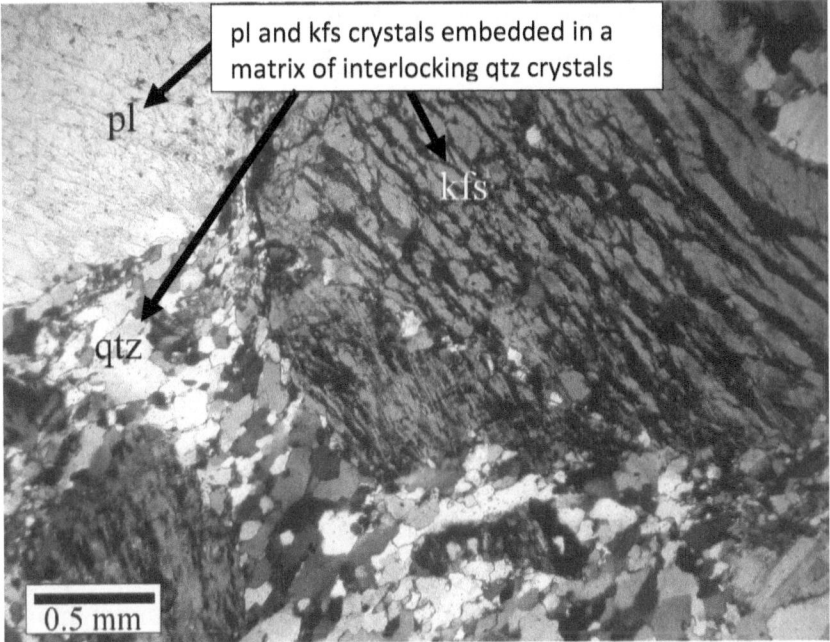

Figure 18.1—A thin section of a granite showing crystals of plagioclase (pl) and potassium feldspar (kfs) embedded in a matrix of quartz (qtz) crystals. Note how the crystals interlock suggesting that "the quartz, at least, was fluid when it was moulded..." as suggested by James Hutton. U.S. Geological Survey file photo.

-----------------------

When Neil Armstrong and Buzz Aldrin landed on the moon in 1969, it captured the imagination of all of America and most of the world. Just the fact that humans could solve the technical problems of flying to the moon was mind boggling in itself. Actually seeing Neil Armstrong reach down and collect his "contingency sample" of moon soil, which he did within minutes of stepping onto the moon, was an unforgettable experience for millions of people watching on television around the world. Astronomers had wondered about what the moon was made of for hundreds of years. More recently, various remote sensing methods had been used to map and describe the rocks on the moon's surface. But with the Apollo 11 mission, it was finally possible to study the rocks of an extraterrestrial body at first hand. And because moon rocks are largely igneous in origin, igneous petrology was a very large part of those studies[3].

When the Neil Armstrong and Buzz Aldrin landed on the Sea of Tranquility on July 20th, 1969, the original flight plan was to let the astronauts rest for four hours before preparing to leave the lunar lander for extra-vehicular activity (EVA). There were two problems with that plan. The first was expecting that Armstrong and Aldrin, after their harrowing landing with less than a minute of fuel remaining, would be in the mood for a quick nap. The second problem had to do with television. They landed at 4:17 PM Eastern Daylight Time and the whole world was watching on TV. Since the checklist for an EVA took about four hours to complete, if they rested for four hours as planned and then started the checklist, they wouldn't begin the EVA until after midnight in the eastern USA. Than meant they would lose a big chunk of their TV audience. So NASA decided to do the EVA immediately, and at 9:56 PM, Neil Armstrong came down the ladder and stepped on the Moon.

Armstrong's first and most important task was to do a visual inspection of the Lunar Module to check for possible damage. Finding none, his next task was to take "contingency samples" of lunar rocks and soil, so that they'd have something to bring back in case the mission had to be aborted. But Armstrong became distracted while taking some photographs, something he decided on his own which was more important. Huston actually had to remind him several times to take the samples, which he finally did.

One of these samples was moon rock number 10022 that Armstrong collected "from the area between the Lunar Module and the flag" (Fig. 18.2).

Figure 18.2—Location where sample 10022 was taken "from the area between the Lunar Module and the flag". NASA file photo.

The rock itself was just three or four inches in length (Fig. 18.3), but just its appearance to the naked eye reveals a lot about how it was formed.

Figure 18.3—One of the "contingency samples" collected by Neil Armstrong. Note the round "vesicles" indicating that this is a volcanic rock that solidified from a lava that contained gas bubbles. NASA file photo.

The rock's most obvious features are that it is composed of fairly dark minerals, and that it contains perfectly round impressions known as "vesicles". Both of those features indicate that this is a basalt, a rock formed by the solidification of lava erupting at land surface, and that it contained a gas phase that formed bubbles. Later, and back on Earth, the geologist/astronaut Harrison Schmitt would describe the rock as "a fine-grained, vesicular, plumose, subophitic, olivine basalt".

That may sound like geologist geek-speak, which it certainly is, but each of those descriptors actually means something specific. The fact that it's "fine-grained" is an indication that the rock solidified from lava erupted at the lunar surface. An erupted lava, whether on Earth or the Moon, cools very quickly and that doesn't give mineral crystals much time to grow. Thus, the fine-grained appearance is good clue as to how it was formed. "Vesicular", as we've already said, means there were gas bubbles in the lava that froze when the liquid basalt cooled. "Plumose" structures are fracture networks that spread outward from the origin of a break in the rock. That, in turn suggests the rock has undergone some trauma—possibly from a meteor impact—after it had solidified. "Subophitic" refers to the texture of the rock and notes that the crystals of plagioclase feldspar (a white mineral of the feldspar family) are approximately the same size as crystals of pyroxene, another indication that the rock solidified quickly. The term "olivine basalt" means that it has a silica ($SiO_2$) content of between 52 and 45 percent, that it is composed mostly of the minerals augite (an iron-bearing black mineral of the pyroxene family) and plagioclase, with traces of olivine. So, if you translate Harrison Schmitt's eight-word description of the rock into English, it would read something like *this is a volcanic rock of basaltic composition that formed from a lava that cooled and solidified very quickly and has been subjected to some kind of a shock during its history.*

The rocks the Apollo missions brought back from the moon have gone a long way toward reconstructing the history of the Moon. What makes that history particularly interesting is that the way the moon was formed is in many ways similar to how the Earth was formed. Both the Earth and Moon condensed as solid bodies out of a nebular dust cloud about 4.6 billion years ago. By 4.55 billion years, the decay of radioactive elements present in the "dust" had largely melted both the Earth and Moon. Gravity then drew the heaviest elements (iron, nickel, and uranium) down toward the core of both bodies[3]. The next heaviest silicate liquids (olivine and pyroxene) settled around the core, and the lightest silicate (a calcium-rich plagioclase known as anorthosite) rose to the moon's surface and formed the first crustal material (Figure 18.4).

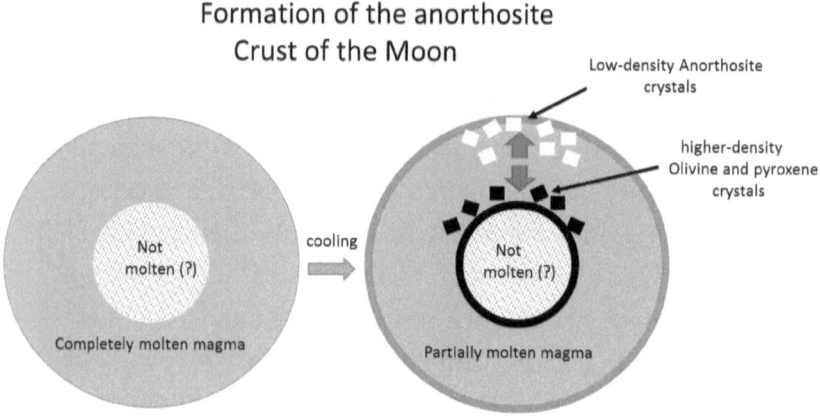

Figure 18.4—Formation of the Moon's anorthosite crust from molten magma.

By 4.5 billion years ago both the Earth and the Moon began to solidify, with a thin crust forming on the outer surface but still with molten liquid underneath. What happened next on the Earth is unknown because no rocks of that age have survived the ravages of plate tectonics and erosion. But on the moon, after it had almost completely solidified, the gravity of the Earth pulled what was left of the molten liquid to the side facing the Earth. As the thin anorthosite crust of the Moon was periodically punctured by asteroid strikes, the underlying molten magma bubbled to the surface where it poured out and pooled in the low spots of the lunar landscape. That, in turn, is what formed the lunar mare, or seas, of which the Sea of Tranquility where Apollo 11 landed is just one. That is how the mare basalt that Neil Armstrong collected, sample 10022, was formed. Sometime after it solidified, sample 10022 was apparently shocked by another asteroid or meteor strike, producing the "plumose" structures that Harrison Schmitt later identified and described.

Then the Moon froze in time. The distribution of rocks below the Moon's surface soils, which were generated by billions of years of meteor impacts and solar radiation, are pretty much the same now as they were after the last mare basalt floods cooled 3.85 billion years ago. That, in turn, has a lot to say about what the Earth might have been like when it cooled to the point that it had a solid crust.

It probably looked a lot like the Moon does today.

But astronauts walking on the moon and collecting rock and soil samples is not the only way to study the igneous petrology of the Moon. The sun is an excellent source of x-rays which are continually bombarding the Moon. When x-rays interact with minerals on the moon's surface, their intensity and characteristic energy levels change, producing a secondary "fluorescent" response. That fluorescence depends on the relative concentrations of some elements, particularly silicon and aluminum, present in the rocks. And, because moon rocks are composed mainly of silicon and aluminum-bearing minerals, that fluorescent response can indicate the kinds of minerals are present. So another way of studying moon rocks is to equip the lunar command module with an x-ray fluorescence detector. That is exactly what was done for the Apollo 15 and 16 missions, and the scientists conducting the experiments were working for the Goddard Space-Flight Center in Maryland[4]. One of those scientists was Dr. Jerry R. Weidner, the University of Maryland igneous petrology professor.

When you look at the moon from earth, the most obvious thing you'll see is that there are distinct differences between the relatively dark "seas", or mare, and the brighter "terra" or highlands. Presumably, those differences reflect something about the composition of the underlying rocks. As the Apollo 15 and 16 Command Modules were orbiting the moon, they continually recorded x-ray fluorescence as well as the brightness, or albedo, of light reflecting off the moon. When the data were transmitted back to earth and analyzed, the scientists quickly noticed that the ratio of aluminum to silicon in the rocks, as indicated by the x-ray fluorescence, correlated almost perfectly with the brightness of the reflected light[4]. Why would that be?

It's because the crust of the moon was initially formed largely from the least dense minerals that floated to the moon's surface as it cooled (Fig. 18.4). One of those low-density minerals was anorthosite that has a 1:1 ratio of aluminum (Al) to silicon (Si) in its crystal structure. The lunar highlands, therefore, have an Al/Si ratio of 1 or higher, depending on what other minerals are present. Furthermore, because anorthosite has a relatively high light reflectance, the highlands appear relatively bright. The mare rocks, on the other hand, were formed from basaltic lavas after the lunar highlands had formed. Those basaltic rocks had a higher proportion of the iron-rich mineral pyroxene which lacks aluminum and which does not reflect light very well. The mare basalts, therefore have a Al/Si ratio closer to 0.5, and they appear as the dark areas on the moon. So, the brightness or lack of brightness, of the different areas we see on the moon are directly related to the relative abundance of minerals like anorthosite and pyroxene. It just goes to show that you don't necessarily have to have a sample of a rock in order to study its igneous petrology.

----------------------

Most geology classes, and igneous petrology was one of these, operated on the theoretical/practical model of teaching. That meant that on Tuesday, Dr. Weidner would lecture on the theory of igneous rock composition and nomenclature. Then, on Thursday the students would do laboratory exercises that addressed the practical aspects of describing, identifying, and classifying igneous rocks. The labs were specifically designed to be done by two students working together as a team. Because the young man and woman had taken calculus and physics together they had gradually become good friends. So, it was natural that they would became lab partners.

One of their first labs had to do with how to measure and describe the texture and mineral composition of igneous rocks. That is because igneous rocks are classified based on those two properties. *Texture* of the rock refers to the size of the crystals that make up the rock. Is the rock composed of fine-grained or coarse-grained crystals? Were the crystals of uniform size or were some crystals larger than others? On the other hand, was the rock composted or contain glass which is not crystalline? Does the rock contain structures like vesicles? And, naturally, there were technical names for all of these textures, and those textures give clues as to the rock's origin. Here are a few examples:

| Property of the rock | Technical name | Possible origin of rock |
|---|---|---|
| Crystals visible to the naked eye | Phaneritic | Plutonic, intrusive rocks |
| Crystals too small for the naked eye | Aphanitic | Volcanic, extrusive rocks |
| Porphyritic | Two crystal sizes in rock. Larger crystals are phenocrysts, smaller are groundmass | Can be found in both intrusive and extrusive rocks |
| Fragmental | Disaggregated igneous material | Pyroclastic rocks from an explosive volcano |
| Pegmatitic | Large crystals > 5 cm | Slowly cooled intrusive rocks |
| glassy | Rapidly cooled lava | Volcanic, extrusive rocks |
| Vesicular | Cavities formed as gases escaped from lava | Volcanic, extrusive rocks |

The other visible property of igneous rocks used in classification is their mineral *composition*. In this respect, igneous rocks are a bit easier to deal with than other types of rocks (sedimentary or metamorphic) because their mineral composition is (usually) simpler. Here are a few examples:

| Rock Composition | Technical Name | Origin of Rock |
|---|---|---|
| Light-colored silicate minerals: 70-55% silica, K-feldspar> 1/3 of feldspars present | *Fel*dspar + *si*lica or **Felsic** | Continental crust |
| Intermediate between felsic and mafic: 65-55% silica, plagioclase feldspars > 2/3 of feldspars present | **Intermediate** | Indeterminate |
| Dark silicate minerals: 50-45% silica, calcium plagioclase predominant feldspar | *Ma*gnesium + *fe*rric iron or **Mafic** | Oceanic crust |

One of the first lab exercises that they had to do consisted of about fifteen different igneous rocks of various textures, compositions, and origins. Their job was simple enough—describe the textures present and measure their mineral compositions. From that they had to assign a rock name and a probable origin of the rock.

The textures and composition of rocks could be observed either in hand sample or by looking at thin sections under a microscope (Fig. 18.1). The most labor-intensive part was determining the rock's composition. That involved identifying the minerals present, measuring their crystal sizes, and then physically counting their numbers in a given area of the rock's surface.

The two lab partners alternated the task of identifying and

counting mineral grains, which was pretty hard work, while the other recorded the data on the lab sheet. The grain-size measurements and tabling the percentages of minerals present took up a lot of space, and the lab sheet eventually grew to more than 10 handwritten pages as they worked through the different rocks. Because this was a fairly tedious process, and because Dr. Weidner would have at least ten labs to grade, the young man began to wonder if he would actually read all of these (boring) descriptions on the lab sheet. So, buried in the middle of one page describing some rock's texture, he slipped in the following hand-written message: *Dr. Weidner, if you read this, see me and collect a soft drink of your choice.*

When the next lecture came around, Dr. Weidner glanced at the young man as he walked into the room. He just smiled and mouthed "diet coke".

## REFERENCES

1. Adams, F.D., 1938. The birth and development of the geological sciences. Dover Publications, Inc., New York, 505 pp.

2. Playfair, J., 1802. Illustrations of the Huttonian theory of the earth: Cadell and Davies. London, England, 528 pp.

3. Smith, J.V., Anderson, A.T., Newton, R.C., Olsen, E.J., Crewe, A.V., Isaacson, M.S., Johnson, D. and Wyllie, P.J., 1970. Petrologic history of the moon inferred from petrography, mineralogy and petrogenesis of Apollo 11 rocks. Geochimica et Cosmochimica Acta Supplement, 1, p.897.

4. Adler, I., Trombka, J.I., Schmadebeck, R., Lowman, P., Blodget, H., Yin, L., Eller, E., Podwysocki, M., Weidner, J.R., Bickel, A.L. and Lum, R.K.L., 1973. Results of the Apollo 15 and 16 X-ray experiment. In Lunar and Planetary Science Conference Proceedings (Vol. 4, p. 2783).

# CHAPTER 19.
## RUNNING SATYR HILL

As the young man approached graduation, his course load had lightened up to the point where he actually had some spare time to fill. His father had discovered running marathons as a hobby a few years earlier—this was in the 1970s at the beginning of what came to be called the "running boom"—and the young man decided to see if he could do it too. After all, if the old man could do it, so could he. The race he decided to train for was the Baltimore Marathon. In those days the race started at Memorial Stadium on 33rd Street, proceeded up Perring Parkway, passed under the Baltimore Beltway, branched off onto Satyr Hill Road which took you to Loch Raven Reservoir, turned around at thirteen miles, and returned along the same route to Memorial Stadium and the finish line.

There were a couple of things that made this a very tough marathon. First of all, the first six miles of the race were all uphill. It was a relatively gentle climb, but still it was uphill and you had to be careful not to go out too fast and exhaust yourself too soon. That was followed by a steep drop from the top of Satyr Hill down underneath the Baltimore Beltway, and then a gentler drop to the turnaround point at Loch Raven Reservoir. Retracing your steps from the turnaround was a gentle climb back toward the Beltway. But then, at about 18 miles out, you returned to Satyr Hill, which was a brutal, steep, one-mile climb.

What made the Satyr Hill climb so difficult was that, at about 18 miles many runners were "hitting the wall", a physiological barrier that wasn't particularly well understood in those days. It turns out that the average human body has enough glycogen—blood sugar that is readily available to support physical activity—to fuel a run of about 18 miles. After that, the body has to switch over to metabolizing fat for energy, which is a less efficient fuel. Hence the feeling of painful exhaustion many recreational runners get around the 18-mile mark of a marathon. In the Baltimore Marathon, the average runner was "hitting the wall" right at the beginning of the Satyr Hill climb—the worst possible time in the run.

But in addition to being a formidable obstacle to marathoners, Satyr Hill is one of several geological oddities that are peculiar to the City of Baltimore. And those oddities reflect a sequence of events that built much of what is now the east coast of North America.

----------------------

The story of how Satyr Hill—the bane of marathoners—became Satyr Hill in the first place begins about 1.1 billion years ago. By that time the various large land masses, or *cratons*, on the Earth had been steadily growing for at least 3.3 billion years. That's a lot of time, and by 1.1 billion years ago the various cratons had grown almost as large as they are today (Figure 19.1). We have to refer to cratons, not continents, since a single modern continent is often composed of more than one craton. Modern South America, for example, is composed of two cratons named Rio de La Plata and Amazonia which were separate 1.1 billion years ago (Fig. 19.1), but have since been welded together. Then as now, continents accrete because of the movements of the tectonic plates in which they are embedded. Whether those plate movements are random or not is still a topic of some debate. However, what we know for sure is that periodically the individual continents tend to clump together and form large supercontinents.

By 1.1 billion years ago, most of the cratons of the world were in the process of moving closer and closer to each other. The red-shaded area of Figure 19.1 shows where plate collisions had turned violent, causing mountain-building events known as orogens. Note that because of the converging plates, a significant orogenic event known as the Grenville Orogeny, was occurring on what was then the southeastern margin of Laurentia, which would later form the core of North America.

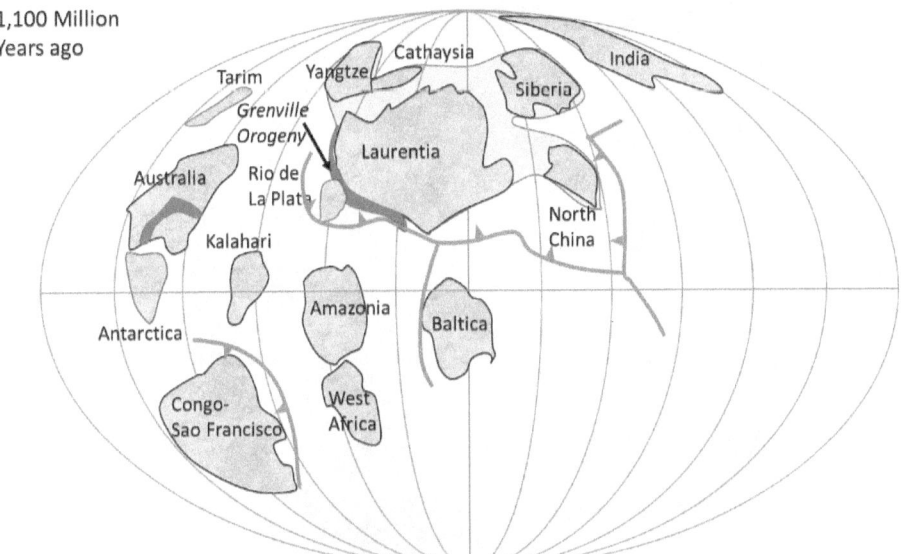

Figure 19.1—Approximate locations of craton land masses 1.1 billion years ago. Adapted from Li and others, 2008[1]. The red-shaded areas represent active orogenic activity and the green lines represent approximate plate boundaries.

Orogenic events are often characterized by lots of volcanic activity, and the 1.1 billion year orogen on the margin of Laurentia was no exception. Volcanos erupted, mountains were built, and those volcanos and mountains were quickly eroded and the resulting sediments were dumped into surrounding ocean. Then, after the sediments were buried, they were trapped between the plate margins and squeezed and heated. The sediments didn't get hot enough to melt, but they did get hot enough to recrystallize into a metamorphic rock called a gneiss that would, in turn, come to be known as the Baltimore Gneiss (Fig. 19.2).

Figure 19.2—A sample of the Baltimore Gneiss, the pink mineral is a potassium feldspar, the dark mineral is biotite, and the light mineral is quartz. U.S. Geological Survey file photo.


For the next couple hundred million years, the continents continued to converge and by 900 million years ago, they had assembled into a super continent known as Rodinia (Fig. 19.3).

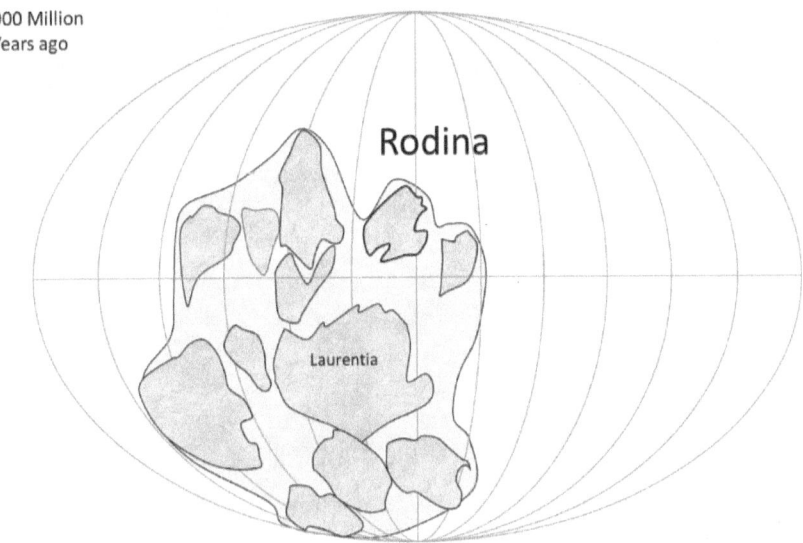

900 Million
Years ago

Figure 19.3—Approximate configuration of the supercontinent Rodinia about 900 million years ago. Adapted from Li and others, 2008[1].

The name Rodinia, incidentally, is taken from a Russian word meaning "The Motherland". When we look at the distribution of the continents that we see today, several things about Rodinia are worth mentioning (Fig. 19.3). First of all Laurentia, which would later become North America, was squarely in the center of the supercontinent. Secondly Amazonia, which would later become part of South America, was located just south of North America, just as it is today. But to the west of Laurentia was Kalahari, which would later become the southern part of Africa. Finally, Laurentia's closest northern neighbor was what eventually would become southern China.

Clearly, the locations of the different continental masses in Rodinia were much different than their locations now, and that reflects the subsequent breakup of the supercontinent. That process took another 400 million years and resulted in the detachment of Laurentia[2] from the other continental masses (Fig.

19.4). That detachment, in turn, is what led to the sediments, later transformed into the metamorphic rocks which now form Satyr Hill in Baltimore.

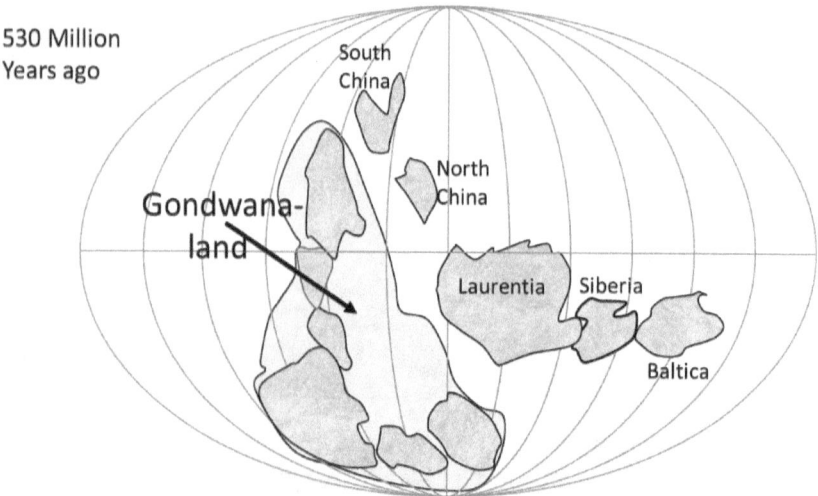

Figure 19.4—The detachment of Laurentia from the supercontinent Gondwanaland 530 million years ago. Adapted from Li and others, 2008[1].

A lot of erosion can occur over the 400 million years that Laurentia was detaching itself from Gondwanaland, and it did. That erosion, in turn, eventually exposed the rocks that had become the Baltimore Gneiss. Interestingly, the erosional surface on top of the Baltimore Gneiss is a world-wide phenomenon that has come to be called the "Great Unconformity". The Great Unconformity was first observed in the mountains of Wales by Adam Segwick (Chapter 11), and also was observed by John Wesley Powell in the Grand Canyon (Chapter 7). The global sea level rise that occurred about six hundred million years ago (because of the melting of the Snowball Earth) drowned the margins of all the cratons including Laurentia, and sediments again began to be deposited. The first sediments that were deposited on top of the remnants of the Baltimore Gneiss were mostly sands that were washing off of Laurentia as it continued to erode. These sands eventually accumulated to a thickness of up to 200 meters, and those sandstones would eventually become what we now call the Setters Formation (Figure 19.5).

But as erosion continued, the sand-producing mountains on Laurentia wore down to the point that little sand or clay was being deposited. That, in turn, allowed the seas adjacent to Laurentia to switch from depositing sands and clays to depositing limestones. The fact that the Cambrian Explosion of shell-bearing organisms had by now occurred (~530 million years) encouraged the production and deposition of carbonate shell material. The resulting limestone which overlays the Setters Formation (Fig. 19.5) was later metamorphosed into the Cockeysville Marble (Chapter 9). After the deposition of the limestone that was to become the Cockeysville Marble, conditions on the continental margin changed again resulting in an influx of more sand and clay sediments. These sediments became what are now called the Wissahickon schist, and they accumulated to a thickness of up to 3,000 meters (Figure 19.5).

After their initial deposition 600-500 million years ago, all of these sediments were eventually buried fairly deeply. That served to turn the sands into sandstones, the silts and clays into shales, and the carbonates into limestone. But repeated collisions with other land masses with the ensuing high pressures and temperatures metamorphosed the rocks into quartzites (The Setters Formation), marbles (Cockeysville Marble), and schists (The Wissahickon schist). Altogether, the Setters Formation, the Cockeysville Marble, and the Wissahickon Formation make up what is now called the Glenarm Supergroup[3]. The generalized stratigraphic succession of these rocks are shown in Figure 19.5.

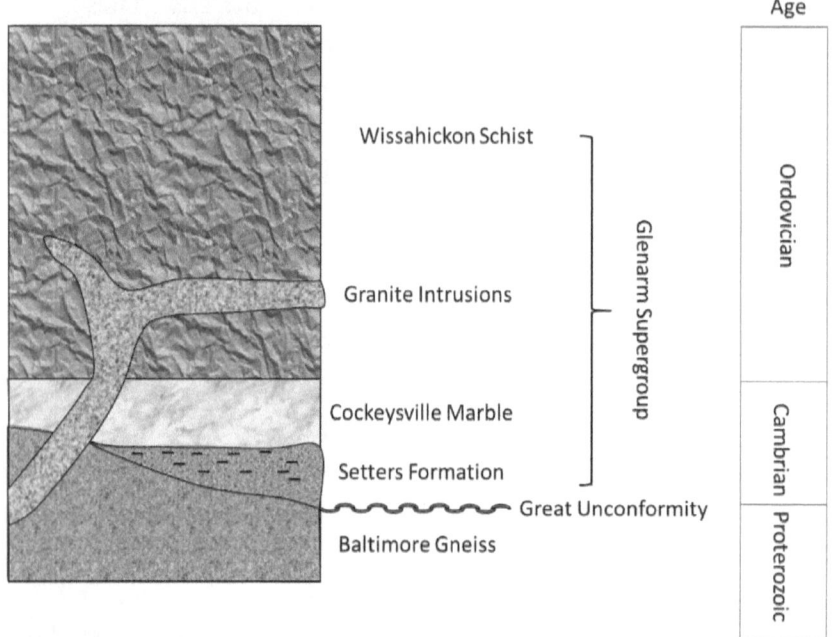

Figure 19.5—Generalized stratigraphy of the Baltimore Complex. Adapted from Muller and Chapin, 1984[3].

As you might expect, all of the collisions with other landmasses did more than just metamorphose the sediments. They also bent, folded, and faulted the rocks that now underlie the city of Baltimore in a variety of ways. These folds and faults are expressed as a series of dome-like structures. The core of these domes is the Baltimore Gneiss. Lapping up on the flanks of the Baltimore Gneiss is the Setters quartzite, followed by the Cockeysville Marble, and then the Wissahickon schist. Each of those formations have been intruded in places by granites that were formed during various stages of the continental collisions.

All of which brings us back to Satyr Hill and the Baltimore Marathon. The Setters Formation, by virtue of being made of a quartz sand that had been welded together during repeated episodes of metamorphism, is a very tough, hard rock. In the humid eastern United States, tough hard sandstones generally form very steep ridges, whereas softer limestones and marbles tend to form valleys. The route of the Baltimore Marathon starts where the Baltimore Gneiss underlies land surface and forms the core of a dome known as the Towson Anticline. The first six miles of the race is a gentle climb up the side of that dome until you reach the hard, tough Setters Formation that forms Satyr Hill (Fig. 19.5). After you go down the steep hill formed by the Setters Formation you run out into the Dulaney Valley which is underlain by the Cockeysville Marble. Finally, when you get to the turnaround near Loch Raven Reservoir, the ground is littered with boulders of the Wissahickon schist that overlies the Cockeysville Marble. Figure 19.6 shows a photo of the young woman sitting near the shore of Loch Raven reservoir amid boulders of the Wissahickon schist.

Figure 19.6—The young woman sitting on a boulder of the Wissahickon schist.

Whether they know it or not, most Baltimorians are intimately familiar with the Setters Formation. Because it is so very hard, and because it cleaves so neatly into regular blocks, it is widely used as a building stone in and around Baltimore. But what really makes it desirable is because while it was being metamorphosed by heat and pressure, long crystals of the mineral tourmaline grew along the bedding planes of the of the original sandstone. Thus, in addition to being a hard rock that naturally breaks into easily usable flat stones, the tourmaline crystals also lend considerable beauty to the stone. Figure 19.7 shows a photograph of the Setters Quartzite with long (~1-3 cm) black tourmaline crystals.

Another unusual feature of the Setters Quartzite is that the tourmaline crystals all seem to be oriented in generally the same direction. That is also a reflection of the manner in which the original sandstone was squeezed and heated during metamorphism. It turns out that when crystals are actively growing during metamorphism, the chemical potential of the atoms that are being assembled are affected by pressure. In the case of tourmaline, the long axis of the mineral (called the c-axis) tends to grow in the direction of the applied stress[4]. The reason the tourmalines in the Setters quartzite line up is because they all grew perpendicular to the direction that they were being squeezed. Furthermore, the reason that the Setters quartzite is such a desirable building stone is a direct reflection of its unusual, and very long, geologic history.

Figure 19.7—The Setters quartzite showing black tourmaline crystals. Note that the crystals are all oriented in the same general direction. Pocket knife is for scale.

---------------------------

The young man had run track and cross-country in high school, but he had never been a particularly fast runner. The great thing about marathons is that you don't have to be fast to be good at it. It just takes a lot of time to put in the training miles in order to build up your endurance. By the time he was ready to run the marathon, the young man was running about 25 miles a week, which is on the low end of what you need to do in order to finish a marathon. But he was, after all, still in college and so 25 miles of training a week was all the time he could spare.

The day of the race in December dawned cold and clear. The temperature at the start was about 25 degrees Fahrenheit, which is actually perfect conditions for a marathon. Heat is the enemy of marathoners and having a relatively cool day to run his first marathon was a good thing. The race started and most of the runners—with the exception of the elite athletes who were actually trying to win the race—went out at a comfortable pace. The young man, knowing he was climbing up the Towson Dome was careful to keep his pace relatively slow, about eight and a half minutes per mile.

The first ten miles or so of a marathon is the easy part of the race, and the runners chat and joke among themselves. Six miles out, the young man came to the top of Satyr Hill and the Setters Formation, and then began dropping down toward the Dulaney Valley and the Cockeysville Marble. By the time he reached the Wissahickon schist at the 13-mile turnaround point, he had been running for an hour and fifty minutes and he was beginning to tire. By now, all of the runners were beginning to feel the stress of fatigue and the jokes and chatter faded away. Everybody was working now.

When the young man reached the base of Satyr Hill, he steeled himself for the climb. But after a quarter of a mile running up the steep face of the hill, he simply ran out of breath. He began walking, as were many of the runners around him. When he reached the crest of the hill at the 20 mile marker, he had caught his breath and actually was feeling pretty good. No evidence of The Wall yet. The young man also knew that the rest of the race was going to be downhill, so he actually picked up his pace. That was a mistake. By the time he reached the 22 mile marker he was seriously tired and he had to slow down. The Wall had struck.

It turns out that running downhill, even down a fairly gentle slope, is very hard on your legs. And after you've run 22 or 23 miles, its murder. Suffice it to say that the last two or three miles where the toughest miles he had ever run. Looking at the faces of the runners around him, he could see they felt the same way. What made it worse was that, about a mile from the finish line, one of the cops who was directing traffic on Perring Parkway turned to a fellow cop and remarked, loud enough for the runners to hear, "these are just the stragglers". Great. Well, by that time the young man was beginning to feel like a straggler.

He crossed the finish line in three hours and 54 minutes, utterly exhausted and with legs that could just barely hold him up. His father had finished before him (that's right, the old man beat him) and jogged the last hundred or so yards with the young man, probably to rub it in. But the young man had finished. He survived.

And he would never forget Satyr Hill.

## REFERENCES

1. Li, Z.X., Bogdanova, S.V., Collins, A.S., Davidson, A., De Waele, B., Ernst, R.E., Fitzsimons, I.C.W., Fuck, R.A., Gladkochub, D.P., Jacobs, J. and Karlstrom, K.E., 2008. Assembly, configuration, and break-up history of Rodinia: a synthesis. Precambrian research, 160(1), pp.179-210.
2. Hoffman, P.F., 1991. Did the breakout of Laurentia turn Gondwanaland inside-out? Science, 252(5011), p.1409.
3. Muller, P.D. and Chapin, D.A., 1984. Tectonic evolution of the Baltimore Gneiss anticlines, Maryland. Geological Society of America Special Papers, 194, pp.127-148.
4. Kamb, W.B., 1959. Theory of preferred crystal orientation developed by crystallization under stress. The Journal of Geology, 67(2), pp.153-170.

# CHAPTER 20.
## CONTINENTS IN COLLISION

One of the last, and in many ways the most demanding,
requirement for a bachelor's degree in geology was Field Camp.
Field camp was where just about everything you had previously
learned about petrology, sedimentology, paleontology, mineralogy,
and structural geology was put to use in order to make geologic
maps. Summer geology field camps are sponsored by many
universities. They are usually six weeks in length and the students
earn six credit hours toward their bachelor's degree. Field camps
are generally designed to be enjoyable as well as being a quality
learning experience, and they usually are. The field camp the
young man signed up for in the summer of 1975 was sponsored by
Virginia Tech in Blacksburg, Virginia. The camp itself, however,
was located in the extreme southwest of Virginia near the little
town of Saltville.

Saltville gets its name from the saltwater springs found
nearby. For thousands of years, this naturally-occurring saltwater
had been used by the Native Americans for flavoring and
preserving meat. When Europeans showed up in the 18$^{th}$ century,
they cooped the salt-producing industry from the local natives and
turned it into a commercial operation. During the Civil War, salt
production from Saltville was an important cog in the Confederate
war effort, and today the town of Saltville (population 2,007 as of
2010) styles itself as the "Salt Capital of the Confederacy" (Fig.
20.1).

Figure 20.1—The sign on VA Route 107 leading into the town of Saltville.

As you might expect, the occurrence of these saltwater springs reflects the local geology[1]. A massive and spectacular system of geologic faults, one of which is named the Saltville Fault, has thrust Cambrian (~530 million years old) limestones and dolomites over top younger Mississippian limestones (~350 million years old). The Mississippian limestones contain beds of rock salt and other minerals formed by the evaporation of ancient seas. The fault-disturbed contact between the Cambrian and Mississippian rocks provides a conduit for groundwater to seep into the salt-bearing limestone, dissolve some it, and to bring it to land surface as springs. While those springs were certainly useful for the Native Americans and later to the Confederacy, the real question is what could possibly have generated the enormous forces necessary to produce those faults in the first place? The radical new theory of plate tectonics was beginning to provide answers to that question when the young man went to field camp in

1975.

The director of the field camp was an experienced field geologist named Dr. Lynn Glover III. He was a quiet unassuming man with a graying beard who had worked for the U.S. Geological Survey for a few years before becoming a professor at Virginia Tech. More importantly, Dr. Glover had been fortunate enough to witness the inception of the theory of plate tectonics while he was a graduate student at Princeton in the 1950s and 60s. In subsequent years, he also made substantial contributions to understanding just what can happen when continents collide.

------------------------

The idea that continents might be able to move around, or "drift", over the surface of the Earth can be traced to a Flemish cartographer named Abraham Ortelius (1527-1598). In 1564 Ortelius published a fairly accurate map of the world showing that the outlines of Africa bore an uncanny resemblance to the outlines of North and South America. In 1587, Ortelius suggested that the Americas had been "torn away from Europe and Africa…by earthquakes and floods", and went on to say that:

> *The vestiges of the rupture reveal themselves, if someone brings forward a map of the world and considers carefully the coasts of the three (continents)."*

The idea of "continental drift" was formally proposed in the modern era by Alfred Wegener in 1912 who based it partly on the "fit" between continents noticed by Ortelius. But in the absence of any apparent mechanism capable of driving continental drift, few geologists believed it at the time. The reaction from American geologists was particularly negative[2]. The British geologist Arthur Holmes suggested in 1919 that convection cells in the Earth's mantle might drive continental drift, but there was no physical evidence that such convection cells were even possible, much less that they actually existed.

Direct evidence for continental drift accumulated slowly in

the early 20<sup>th</sup> century beginning with the work of Harry H. Hess, who became a professor of geology at Princeton University in the 1930s. Hess was looking at the problem of continental drift from a somewhat different perspective than Wegner or Holmes. He was not so much interested in continents moving around as he was in the geology of ocean basins. In the 1930s, the ocean basins of the world were considered by most geologists to be very ancient and unchanging features of the earth. Hess' work was instrumental in turning that conventional thinking on its head.

For a variety of reasons, the U.S. Navy had an interest in making gravity measurements of oceanic seafloors, and equipped a submarine with a towed gravimeter to make them. Hess wanted to participate in that research, but he was required to take a commission in the Naval Reserve in order to serve aboard the submarine. This new instrumentation soon made it apparent, at least to Hess, that the geology of the seafloor was much more complicated than was previously thought. However, when World War II started, those studies were put on hold and Hess was called into active duty with the Navy. He subsequently became the captain of an attack transport ship named the *USS Cape Johnson*, and was deployed to the Pacific theater. But rather than interrupting his research, that assignment gave Hess the opportunity of a lifetime.

As anybody who's ever been in the navy can attest, long voyages to distant places can be excruciatingly dull and boring. Fortunately for Hess, however, the *Cape Johnson* was equipped with a sonar for making soundings of the seafloor. But instead of just using the sonar to safely navigate shallow waters, which is what it was usually used for, Hess simply left the instrument on continuously as he was crossing the Pacific Ocean. Hess did this out of simple curiosity and because, well, there wasn't much else to do. What he found was that the ocean bed was not just an empty bowl-shaped "floor" like most geologists thought. Rather it had a distinct topography that included "rises" where the water was shallower in places, trenches that were incredibly deep, and contained flat-topped mounts that he named Guyots (after a fellow Princeton geology professor). All of this suggested that, rather than being a stable, quiescent, non-changing environment, the seafloor was a dynamic place where a lot was going on.

Hess' ship fought in four major engagements during the war including the Battle of Leyte Bay in the Philippines, the largest the largest naval battle in history. But when the war was over, Hess began publishing the results of the studies he had done while traversing the Pacific Ocean. In addition, he began a Caribbean Research Project that was supported by the Office of Naval Research, the National Science Foundation, and several private oil companies. By 1973, that research project had resulted in a total of thirty-four Ph.D. dissertations, one of which was the dissertation of Lynn Glover III, who was to become the young man's professor at Virginia Tech's geology field camp.

In 1962, Hess published a paper entitled *History of Ocean Basins*[3] in which he laid out evidence that ocean basins were (1) underlain by relatively young basaltic rocks, (2) those basaltic rocks were derived from lavas originating in the Earth's mantle, (3) some ocean basins were continually expanding at mid-oceanic ridges, and (4) the rate of ocean-floor expansion at the Mid-Atlantic Ridge was on the order of 1 centimeter per year. That suggestion was confirmed in 1963 when Frederick Vine and D.H. Mathews showed that the magnetic polarization of the basalts on either side of the Mid-Atlantic Ridge had reversed periodically over the last few million years, and that the patterns, or stripes, of polarization on each side of the ridge were virtually identical[4]. All of this showed conclusively that the Atlantic Ocean basin is, and has been for some time, expanding. That, in turn, suggested that continents embedded in the earth's crust could be moved around by the mechanism of sea-floor spreading.

That possibility presented opportunities for field geologists like Lynn Glover. If sea floor spreading was opening up new ocean basins, then it was possible that continents could be moving as well, causing them to periodically collide. What did that imply about the origin of the folding and faulting that were clearly visible in Appalachian Mountains? Much of the Appalachians had been mapped previously, but it was often done at a very broad scale and was not particularly well-suited for identifying continental collisions, when they might have occurred, and what kinds of deformation they caused. In 1973, Glover had this to say about the status of geologic mapping in the Appalachians[5]:

> Before any evolutionary model (of the
> Appalachians) can stand with confidence, we must
> know more about the geologic framework of the
> central and southern Appalachian Piedmont (of
> which) less than 15 percent of the piedmont is
> covered by modern detailed mapping at a scale of
> 1:62,500.

So, in the 1970s, Lynn Glover, and many other geologists, fanned out over the Appalachian Mountains to fill in those gaps.

The young man and the other students at the Virginia Tech geology field camp were familiar with the theory of plate tectonics. But theory is just theory. In the first week of their fieldwork they were given a ring-side seat for just what continental collisions could really do. The Saltville Fault was a case in point[1]. That particular kind of fault, called a *thrust fault*, which simply means that one body of rock has been "thrust" over top of another. That can only happen if compressive force, a *lot* of compressive force, is applied to them. In 1975, plate tectonic theory was suggesting that those compressive forces were the result of continents and/or ocean basins colliding with each other.

The first project the student were given was to measure the thickness of a section of Paleozoic rocks exposed to the east of Saltville along Route 107. The rocks of the Honaker Dolomite of Cambrian age were clearly visible just above the Saltville fault. The first thing the students noticed was that the gray-colored

Honaker Dolomite had obviously been fractured and shattered. Those cracks and fractures had subsequently been filled with white calcite cement over the millennia, which made them stand out clearly against the grey dolomite. It was pretty obvious, even to the inexperienced students, that a tremendous amount of force had been required to thrust those rocks, shattering them in the process. Just how much force, however, when that force had been applied, and from what direction the force had come was not immediately obvious. But in the summer of 1975, those were precisely the questions that Lynn Glover was trying to answer.

-----------------------------------

Glover's first foray into this new and exciting research was near the little town of Virgilia, Virginia, which was close to the North Carolina State Line. Detailed mapping revealed that the area was underlain by a series of volcanic rocks and sediments eroded from volcanic rocks. Furthermore, these rocks had been compressed and folded with an axis that suggested the compression had come from the east. As the deformation had continued over time, the rocks were intruded with an igneous granodiorite, a granite that contains an abundance of dark minerals. Because zircon crystals had formed as both the volcanic rocks and the granodiorite were solidifying, it was possible to date the beginning (the volcanic rocks) and the end (the granodiorite) of the compression event that formed them. The resulting $Pb^{206}/U^{238}$ and $Pb^{207}/U^{235}$ dates indicated that the compression began about 620 million years ago, and ended about 575 million years ago. Glover named this event the *Virgilina Deformation*[5].

Glover published these results in 1973. But other geologists were also working on the same problem—what compression events had occurred in the past and when did they occur? Furthermore, they were seeing patterns of deformation in other places that were similar to what Glover had seen in Virgilina. Although there were (spirited) disagreements over the details, a general consensus emerged that a series of collisions had occurred between oceanic crust to the east and the ancestral North America to the west[6] beginning in late Precambrian time (620 million years ago). As the oceanic crust dived under North America, a volcanic island arc developed offshore with the resulting sediments eventually being squeezed and heated between converging continental and oceanic crust (Figure 20.2).

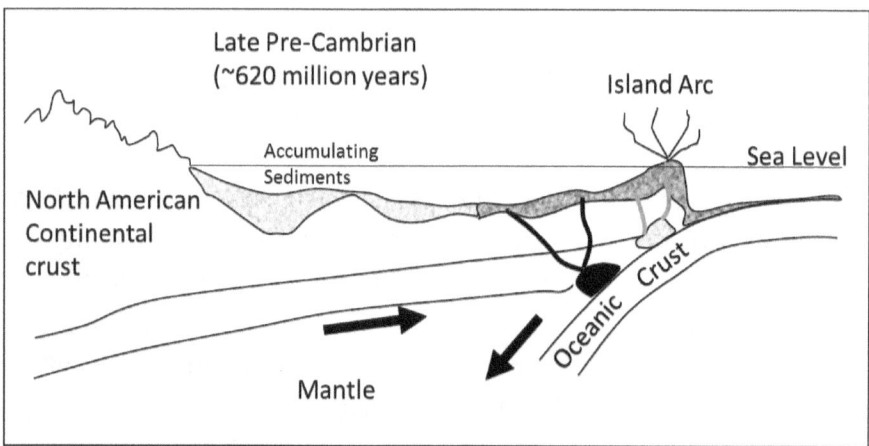

Figure 20.2—Collision between the ancestral North America and oceanic crust causing compression and the formation of an island arc in the late Precambrian (620-575 million years). Adapted from Hatcher, 1972[6].

But the collisions didn't end in the late Precambrian. As it happens, the events that caused the *Virgilina Deformation* were just the beginning of a long and involved history of collision and deformation that went on by another 200 million years. By 1983, geologic mapping and dating of deformation events had progressed to the point that Glover his colleagues could begin to piece together the remarkable and violent history of the Appalachian Mountains[7].

The initial collisions between the spreading oceanic crust diving beneath North America, of which the *Virgilina Deformation* was just one example, are often lumped together in to what is called the *Taconic Orogeny*, named after the Taconic Mountains in

New York. The Taconic orogeny ended about 440 million years ago, but was immediately followed by another volcanic arc known as *Avalonia*, again formed because of a spreading ocean floor to the east of North America (Fig. 20.3 A) about 420 million years. By 370 million years, the Avalon terrane had slammed into North America causing another mountain building event known as the Acadian Orogeny. That, incidentally, was what finally delivered The Rock of Mattapan to what is now New England (Chapter 3). By now, the super continent of Gondwanaland was approaching North America, shrinking the intervening ocean as it came (Figure 20.3 B).

By 320 million years ago, North America and Gondwanaland had collided, forming the supercontinent of Pangea (Figure 20.3 C) and precipitating the Alleghenian orogeny. The Alleghenian orogeny ground on until about 290 million years ago (Figure 20.3 D). These series of collisions buckled, folded, and faulted rocks up and down what are now the eastern United States, leaving the dizzying array of folded and thrust-faulted mountain ranges that we now call the Appalachians.

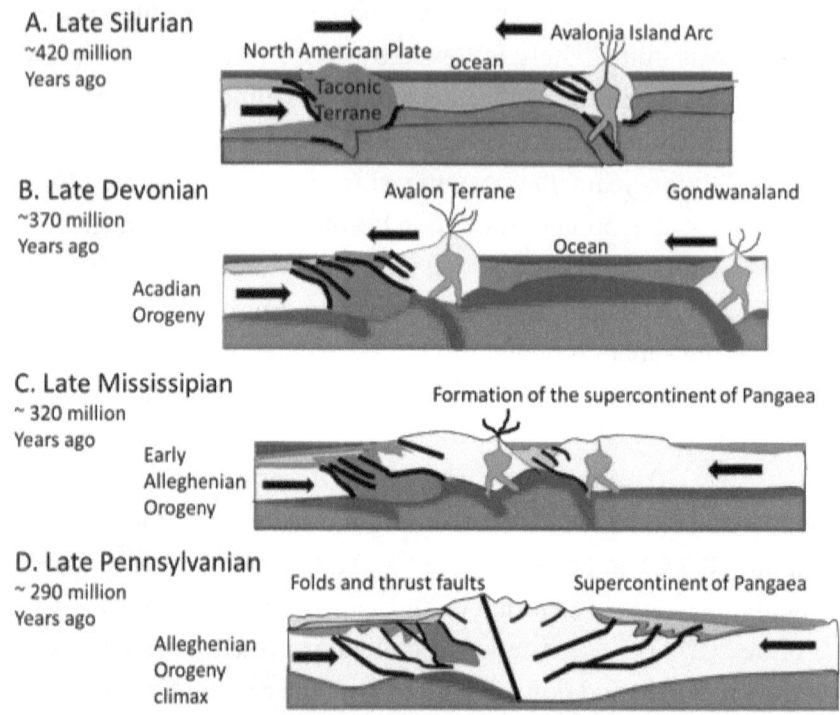

Figure 20.3—Development of the Appalachian Mountains during the Acadian and Alleghenian orogenies from 420 to 290 million years ago. Source: U.S. Geological Survey[8].

---------------------------------------

The young man's immediate problem that summer was to complete measuring the section along Route 107 leading east out of Saltville. At the bottom of the section, starting at the Saltville Fault, was the Honaker Dolomite, which had been deposited in the late Cambrian seas that flooded the edges of North America west of the Taconic terrane (Fig. 20.3 A). Over the next few days, he and his field partner (the center-fielder for Clemson University's baseball team) worked their way up section describing the different rocks they encountered and measuring their thickness. That included the Honaker Dolomite, which graded up into the limestones of the Nolichucky Formation, also of Cambrian age. The Nolichucky limestone was notable because it contained

numerous traces of worm tracks on the bedding planes. The young
man wasn't impressed by those worm trails at the time, but they
were evidence that life on earth was shifting from being dominated
by single-celled microorganisms to more advanced multicellular
creatures, including worms (Chapter 14).

As they progressed upward in the section, they came across
limestones comprised of bits of broken up crinoids and bryozoans,
evidence they had passed through the Cambrian and into the
Ordovician. Continuing up section they encountered the
Martinsburg shale, the Juniata and Clinch Formation sandstones of
late Ordovician age, the and at the top of the section they passed
through the Milboro Shale of Devonian age and finished the
section at the Brallier Shale, also of Devonian age. By the time
they finished measuring the section, The young man and his
partner had traversed some 140 million years of Earth's history in
a little less than a week of fieldwork. A simplified version of the
Route 107 section they measured is shown in Figure 20.4.

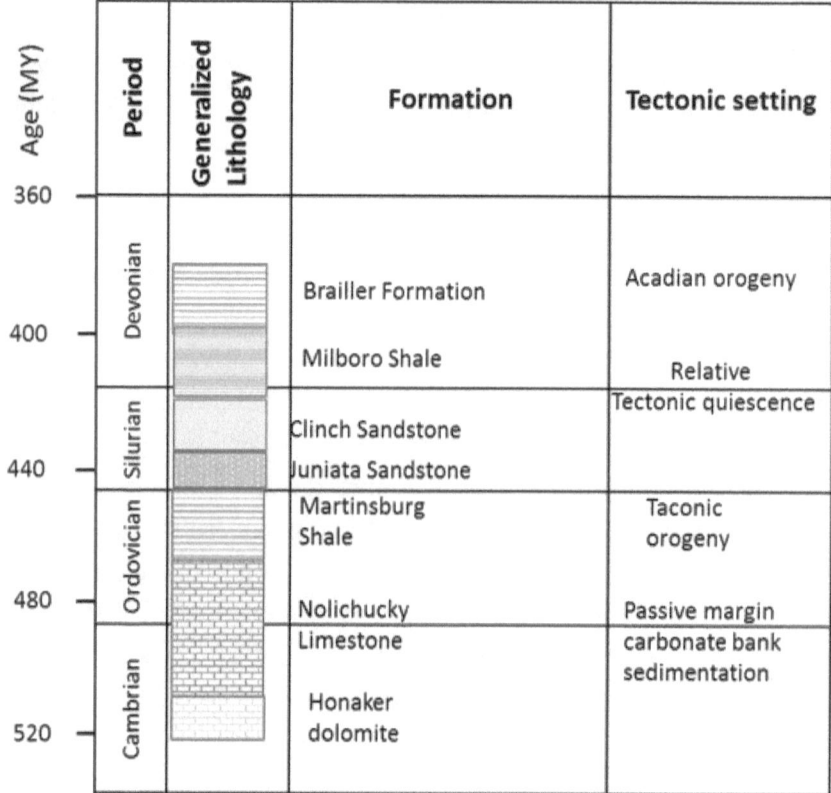

| Age (MY) | Period | Generalized Lithology | Formation | Tectonic setting |
|---|---|---|---|---|
| 360 | | | | |
| 400 | Devonian | | Brailler Formation | Acadian orogeny |
| | | | Milboro Shale | Relative |
| 440 | Silurian | | Clinch Sandstone | Tectonic quiescence |
| | | | Juniata Sandstone | |
| 480 | Ordovician | | Martinsburg Shale | Taconic orogeny |
| | | | Nolichucky | Passive margin |
| 520 | Cambrian | | Limestone | carbonate bank sedimentation |
| | | | Honaker dolomite | |

Figure 20.4—A simplified rendering of the measured section along VA Route 107. The interpreted tectonic settings are taken from Read and Eriksson (2016)[9].

As they worked, the young man and his partner had focused mostly on what the rocks were (dolomite, limestone, sandstone, shale) and hadn't thought too much about what that succession might mean. But as they were finishing up at the end of the week, they were visited by another Virginia Tech Professor named Dr. J. Fred Read. Read was busy with his own fieldwork, but he took the time to walk part of the section with the little crowd of students, and introduce them to what the rocks might mean in terms of tectonic plate movements and collisions.

The limestones and dolomites, according to Dr. Read[9], recorded a time when the seas rose and flooded the margins of a heavily eroded North America continent during Cambrian time. Those seas were relatively shallow, and were dominated by carbonate shelf sediments (the Honaker dolomite and Nolichucky limestones). But as a spreading seafloor began to dive under North America, a deep oceanic trench developed (Martinsburg shale) followed by the rising mountain ranges of the Acadian Orogeny that delivered sands (Juniata and Clinch sandstones) followed by development of a deep-water gray-black shale (Milboro shale), and finally a broad clayey delta (Brallier shale).

Dr. Read's point, which was a revelation to the students, was that the various kinds of rock they encountered as they worked up the section were not randomly distributed. Rather, they were a direct reflection of what was going on in terms of the plate tectonics at the time they were deposited. It gradually began to dawn on the students that the real importance of identifying and measuring the thickness of different kinds of rocks is not so much to describe what they *are*, but to understand what they *meant*.

They were beginning to think like geologists.

## REFERENCES

1. Whisonant, R.C. and Radford, V.A., 1996. Geology and the Civil War in southwestern Virginia: the Smyth County salt works. Virginia Minerals, 42.

2. Adams, F.D., 1938. The birth and development of the geological sciences. Dover Publications, Inc., New York, 505 pp.

3. Hess, H.H., 1962. History of Ocean Basins. Petrological Studies: A Volume in Honor of AF Buddington. AEJ Engel, et al, 599.

4. Vine, F.J. and Matthews, D.H., 1963. Magnetic anomalies over oceanic ridges. Nature, 199(4897), pp.947-949.

5. Glover III, L. and Sinha, A.K., 1973. The Virgilina deformation, a late Precambrian to Early Cambrian (?) orogenic event in the central Piedmont of Virginia and North Carolina. American

Journal of Science, 273, pp.234-251.

6. Hatcher, R.D., 1972. Developmental model for the southern Appalachians. Geological Society of America Bulletin, 83(9), pp.2735-2760.

7. Glover III, L., Speer, A., Russell, G.S. and Farrar, S.S., 1983. Ages of regional metamorphism and ductile deformation in the central and southern Appalachians. Lithos, 16(3), pp.223-245.

8. Reed, J.F., and Eriksson, K.A., 2016. Paleozoic sedimentary successions of the Virginia Valley & Ridge and Plateau, *in* Bailey, C.M., Sherwood, W.C., and Powars, D.S., eds., Geology of Virginia: Virginia Museum of Natural History Special Publication 18, p. 17-54.

# CHAPTER 21.
## A DISTURBED PUZZLE

In the spring of 1976 the young woman registered for a field camp conducted by Eastern Illinois University. The actual location of the camp, however, was based in Rapid City, South Dakota. The dormitories of the South Dakota School of Mines were largely empty during the summer, and that's where the students and teachers would be staying. This was a great place for a geology field camp for a couple of reasons. First and foremost, the geology of western South Dakota is wonderfully diverse, with sedimentary, metamorphic, and igneous rocks occurring in a variety of settings and ages. But also, by virtue of South Dakota's arid climate, the rocks are much more exposed than in, say, Illinois. That made it easier for the inexperienced students to actually see the rocks they were mapping. So, in the middle of June, the young woman packed up her field clothes and boots and headed west to South Dakota.

The first project that the students were assigned was to map a portion of the Badlands National Park which is south and east of Rapid City. The Badlands are visually very striking and quite beautiful (Fig. 21.1). But for students just beginning to learn the art of geologic mapping, they have the very great virtue that you can easily see the contacts between the different lithologic units and literally follow them to the horizon. That meant that if you measured just a few sections of rock in some detail, you could visually extrapolate the contacts between the units all the way to the horizon.

Figure 21.1. The young woman in the Badlands of South Dakota. The highly visible stratigraphic succession is ideal for undergraduates learning how to make geologic maps.

A stratigraphic column of part of the Badlands is shown in Fig. 21.2 and consists of the Pierre (pronounced "Peer") Shale and the Fox Hills Formation[1]. The Pierre Shale is largely grey in color and was deposited by the shallow sea that covered much of western North America 70 million years ago. It is characterized by marine fossils, notably an extinct genus of cephalopods known as *Baculites* that have a distinctive straight shell. As time went on during the deposition of the Pierre Shale, the depth of the water overlying what would become the Badlands grew progressively shallower. By the time the Fox Hills Formation was being deposited, the sea had partially withdrawn and the sediments were deposited by a series of deltas on the edge of the sea. The distinctive yellow and red colors characteristic of the Fox Hills sediments are caused by oxidized iron coatings on the sediment grains, which in turn reflects deposition by fresher rather than more saline water.

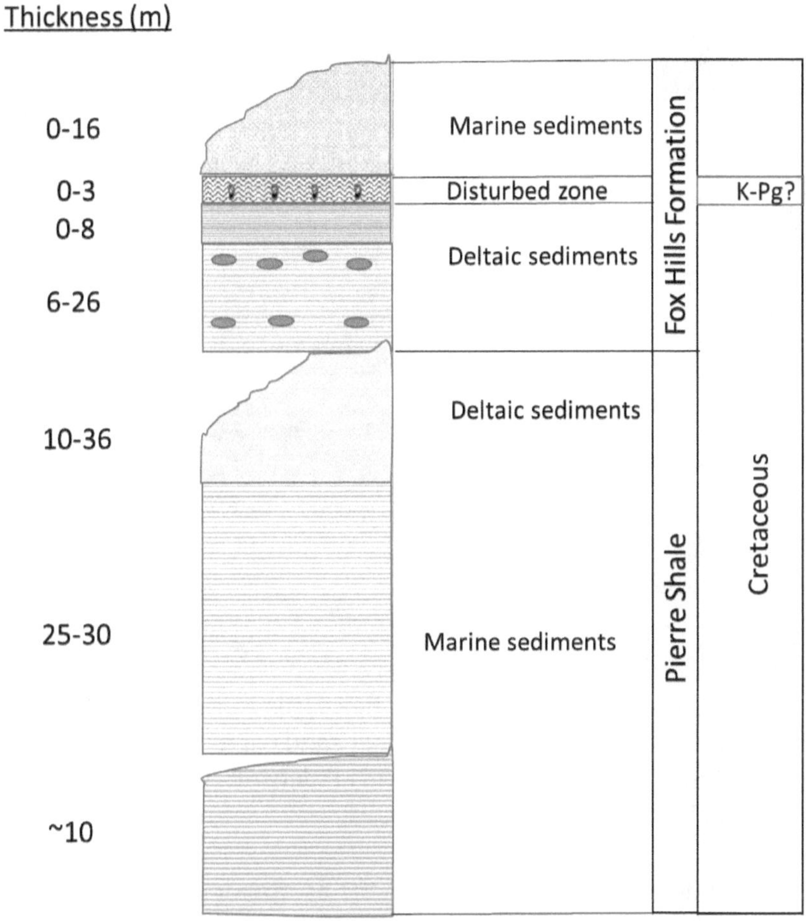

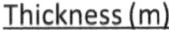
Thickness (m)

| 0-16 | Marine sediments | Fox Hills Formation | K-Pg? |
|---|---|---|---|
| 0-3 | Disturbed zone | | |
| 0-8 | Deltaic sediments | | |
| 6-26 | | | |
| 10-36 | Deltaic sediments | Pierre Shale | Cretaceous |
| 25-30 | Marine sediments | | |
| ~10 | | | |

Figure 21.2—Cretaceous and lower Paleogene stratigraphy of Badlands National Park (North Unit) adapted from Stoffer, 2003[1].

One afternoon, the three students in the young woman's field team were measuring a section of the Fox Hills Formation (Fig. 21.3) when one of the professors came up to watch them work.

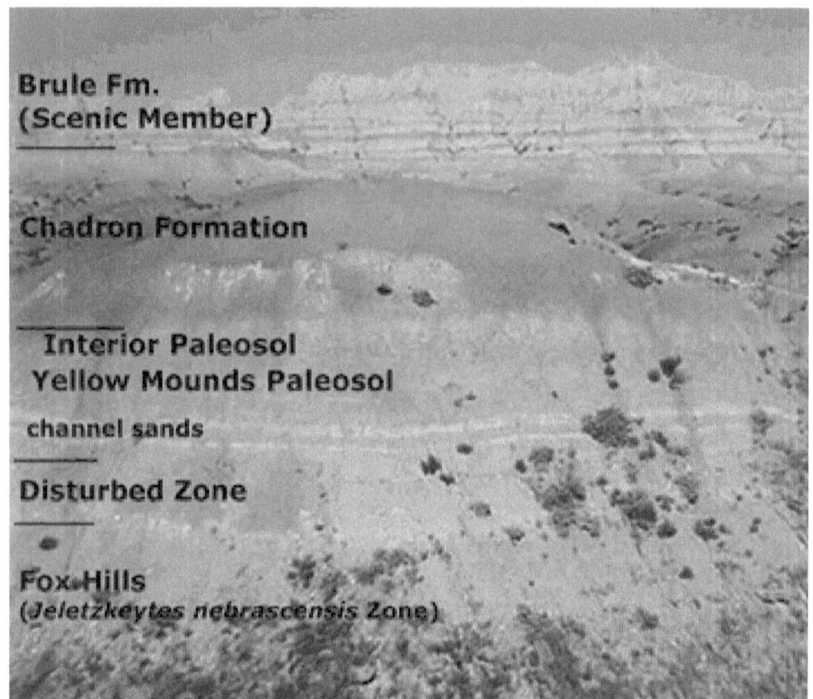

Figure 20.3—Outcrop showing the Fox Hills Formation grading upward to a "disturbed zone", channel sands, and an ancient soil horizon known as the Yellow Mounds Paleosol. Photo by the U.S. Geological Survey (Stoffer, 2003[1]).

As the students progressed upward, the professor pointed to an unusual-looking layer about 1½ meters thick, and asked them what they thought it was. The young woman and her two field partners paused and took a closer look. The sediments were distinctly different in appearance from the underlying and overlying beds. For one thing, the zone didn't seem have any obvious bedding planes like the sediments above or below it. Some of the sediments looked like they had slumped and there were little bits of charcoal embedded in them. How could it be, the professor wondered aloud, that these deltaic sediments could have been deposited without showing any obvious bedding planes?

The students stared, not knowing what to think. Finally, one of them asked the professor what he thought. The professor just shrugged. "No idea", he said "I've wondered about that zone for years and can't figure it out. You can trace it over most of the Badlands Park, but as far as I know, nobody knows how it was formed. It's just a puzzle that we don't have an answer for".

Figuring out how rocks came to be the way they are is a lot like trying to solve a huge jigsaw puzzle. The difference is that with most jigsaw puzzles you already know what the picture you're assembling looks like. That way, you can work backwards in order to figure out how the pieces fit together. But with geology, it works the other way around. You have to fit the pieces together first in order to see the picture. The "pieces" you work with are observations of what the rocks look like, the structures that they contain, and how they relate to each other.

So, as the students and the professor picked through the sediments of the strange zone, they considered the few pieces of the puzzle that they could see. The lack of visible bedding planes had to be one clue. Could that mean the zone represented something like a mudslide? That would also explain the apparent slumping. But how could a mudslide explain the fact that the zone was so widespread? Slump deposits are typically fairly localized. Also, what could the fragments of charcoal mean? After thinking it over it for a little while, they gave up and continued measuring up the section. There simply weren't enough "pieces" to figure out what the picture was. In their notes, they simply called it "a zone of unknown origin".

Then they moved on.

------------------------------

One early description of the Fox Hills Formation, by N.H. Darton of the U.S. Geological Survey in 1905[2], is decidedly unremarkable:

> In some districts, the Fox Hills beds begin abruptly
> with a sudden change from the dark shales of the
> Pierre to sandstones or sandy shales of the Fox
> Hills containing some distinctive species.

A later description by W.V. Searight in 1937 is similarly unremarkable[3]:

> *The contact between Pierre and Fox Hills is one of transition from shale through sandy shale to fine sandstone of the Fox Hills.*

But because the Fox Hills Formation in the badlands is so well exposed, and because the National Park makes those rocks accessible for study, geologists like the young woman's field camp professor couldn't help but notice oddities like the "zone of unknown origin". That zone clearly recorded something out of the ordinary that happened during or after sediment deposition. But what?

It wasn't until 1980 that some more pieces of the puzzle began to emerge. That was when Luis and Walter Alvarez first suggested that an asteroid had hit the earth at the end of the Cretaceous (K) period and beginning of the Paleogene (Pg) period (Figure 11.1), effectively killing off the dinosaurs[4]. That theory generated a lot of interest, and suddenly geologists the world over began looking for evidence of an asteroid impact at the K-Pg boundary[5]. In 1983, three years after the Alvarez paper was published, a geologist named Greg Retallack provided stratigraphic evidence that the K-Pg boundary might be present in "deeply weathered" sediments of the Fox Hills Formation in the Badlands Park.[6] Could it be that the obviously disturbed "zone of unknown origin" that The young woman's field camp professor had pointed out might have been caused by the K-Pg asteroid impact?

That possibility was considered by Phillip Stoffer and his associates at the U.S. Geological Survey and Temple University. They made a detailed study of the Badlands geology, coining the term "disturbed zone" for the strange-looking bed[7]. The disturbed zone sediments for the most part lacked visible bedding planes. But in places, they found isolated chunks of sediment a few centimeters in size that actually contained visible stratification. That implied the sediments had had bedding planes when first deposited, but somehow they had been shaken or somehow jumbled up. They also found spiral-like sediment structures that looked for all the world like rolled up rugs. Furthermore, those roll structures were oriented in a consistent east-west direction. Could those sediment structures have been rolled up by a shock wave? Finally, in addition to the bits of charcoal visible to the eye, they found microscopic spherules of melted sediment and grains of shocked quartz typically associated with meteor impacts. All of these newly discovered clues suggested that some sort of fiery conflagration had struck the sediments soon after they were deposited. Could that conflagration have been the shock wave produced by the K-Pg asteroid impact that occurred 2,000 miles to the south? Initially, that is what the team concluded[8].

But if indeed it was the shock wave of the K-Pg asteroid that formed the disturbed zone, then that would mean that any sediments laying above the zone must be Paleogene or younger in age. For the next several years the team focused on getting reliable age dates for the Fox Hills sediments above and below the disturbed zone.

That wasn't easy. It is possible to date fossilized shell material using ratios of radiogenic strontium ($^{87}Sr$) and its daughter product ($^{86}Sr$). Fossil ammonites found immediately below the disturbed zone yielded an average age of 67.6 million years ago. The K-Pg boundary is generally accepted to slightly younger (66 million years) and so that was consistent with the idea that the disturbed zone was indeed at or near the K-Pg boundary. Unfortunately, no fossils visible to the naked eye were to be found *above* the disturbed zone, so strontium age dates couldn't confirm that the disturbed zone actually marked the K-Pg boundary.

While the sediments within and just above the disturbed zone lacked visible fossils, they did contain microfossils called dinoflagellates. Those microfossils indicated an age date older than the K-Pg boundary by one or two million years. That, in turn, suggested that the disturbed zone wasn't the result of the K-Pg asteroid impact after all[9].

But meteors and asteroids are hitting the earth all the time. Just because it might not be the K-Pg asteroid that produced the disturbed zone didn't mean it couldn't have been caused by a different, later meteor impact. As it happens, debris from another impact event was discovered in Colorado in 2003 that has been dated to 68 million years—the same age as the Badlands disturbed zone[9]. Could that have been the cause of the disturbed zone? So, even though many more pieces of the jigsaw puzzle had been assembled by 2008, the picture it showed was still not entirely clear. [9]

None of this could possibly occurred to the young woman, to her field-camp mates, or to their professor that day in Badlands National Park. They saw the jumbled nature of the bed, they saw the charcoal bits, and they could plainly see that something unusual had happened. But in 1976, the idea that meteors and asteroids could slam into the earth killing off large swaths of life on earth and producing beds of hopelessly disturbed sediments was beyond their youthful imaginations.

-------------------------

As the six weeks of the field camp unfolded, The young woman and her fellow students completed the mapping project in the Badlands National Park and moved on to other things. Those projects included making geologic maps of parts of the Black Hills and of Bear Butte State Park. The Black Hills are formed of ancient (1.8 billion year-old) granites, pegmatites, and gneisses. That, in turn gave the students some experience in mapping igneous and metamorphic rocks. The famous gold mineralization in the Black Hills was emplaced more recently when the Rocky Mountains began being raised about the same time as the K-Pg asteroid impact. The young woman and the students actually visited the Homestake Gold mine, which operated from 1877 to 2002, and was still producing gold in 1976. She only remembers that it was a really hot day, that the machinery was dreadfully loud, and that the arsenic pits used to extract gold from the crushed ore smelled terrible.

But the young woman never forgot the experience of puzzling over the origin of the disturbed zone she'd seen in the Badlands. She, her professor, and the other students knew it meant *something*, but what? Years later, when she read about the possible asteroid impact(s), she just nodded her head.

The "disturbed zone" had indeed been disturbed after all.

## REFERENCES

1. Stoffer, P.W., 2003. Geology of Badlands National Park: a preliminary report (pp. 1-63). US Department of the Interior, US Geological Survey Open-File Report 03-53 63-pp.
2. Darton, N.H., 1905. Preliminary report on the geology and underground water resources of the central Great Plains. U.S. Geological Survey Professional Paper no. 32. US Government Printing Office.
3. Searight, W.V., 1937. Lithologic stratigraphy of the Pierre Formation of the Missouri Valley in South Dakota: South Dakota Geological Survey Report of Investigations no. 27.
4. Alvarez LW, Alvarez W, Asaro F, Michel HV, 1980. Extraterrestrial cause for the Cretaceous–Tertiary extinction. Science. 208 (4448): 1095–1108.
5. Schulte, P., Alegret, L., Arenillas, I., Arz, J.A., Barton, P.J.,

Bown, P.R., Bralower, T.J., Christeson, G.L., Claeys, P., Cockell, C.S. and Collins, G.S., 2010. The Chicxulub asteroid impact and mass extinction at the Cretaceous-Paleogene boundary. Science, 327(5970), pp.1214-1218.

6. Retallack, G.J., 1983. Late Eocene and Oligocene paleosols from Badlands National Park, South Dakota. Geological Society of America Special Paper 193.

7. Stoffer, P.W., Messina, P., Chamberlain Jr, J.A. and Terry, D.O., 2001. The Cretaceous-Tertiary Boundary Interval in Badlands National Park, South Dakota. US Geological Survey Open-File Report, 49 pp.

8. Terry, D.O., Chamberlain, J.A., Stoffer, P.W., Messina, P. and Jannett, P.A., 2001. Marine Cretaceous-Tertiary boundary section in southwestern South Dakota. Geology, 29(11), pp.1055-1058.

9. Jannett, P.A. and Terry, D.O., 2008. Stratigraphic expression of a regionally extensive impactite within the Upper Cretaceous Fox Hills Formation of southwestern South Dakota. Geological Society of America Special Paper 437, p.199-213.

# CHAPTER 22.
## POETS, BOULDERS, AND GLACIERS

The young man and woman were married in 1979, shortly after they graduated from the University of Maryland. While the Arab oil embargo of 1973-74 had been traumatic for America, it also set off a boom in oil drilling activity and geologists were in demand throughout the 1970s. But there were also lots of opportunities for geologists that didn't involve drilling for oil. The young woman went to work for an engineering company, Century Engineering, which had a geotechnical division that supported and inspected construction projects. The young man, thanks to a recommendation from one of his professors, landed a job with the U.S. Geological Survey in what was then called the Water Resources Division.

They had to move to Baltimore, which for the young woman meant moving away from home, but it was worth it. With both of them working full-time jobs, they suddenly had something that both had lacked during their student days: Money. In later years, with house payments and kids to raise, money would again become an annoying limitation. But for those first magical years of marriage, money ceased to be an issue. Which meant that, after a year and a half of working their new jobs, they could actually afford to take a real vacation. The young woman's sister had gone to college in Edmonton, Alberta, where several of their cousins lived. The young woman had never taken an extended road trip to see the country, and so it was decided that that summer they would drive the 2,000 miles to Edmonton to visit her sister. From Edmonton, they'd visit the Canadian Rockies to the west and take time to see the towns of Banff and Jasper.

They left in the morning in early June, taking the Pennsylvania Turnpike over the Appalachian Mountains to the Ohio Turnpike, and then west towards Chicago. On the Ohio Turnpike they stopped a couple of times to collect fossils from the limestone outcrops they noticed near Sandusky, Ohio. The limestones contained an abundance of rugose corals, probably of Devonian age, that are often called "horn corals" because, well, they look like miniature horns (Figure 22.1).

Figure 22.1—Rugose corals, probably of Devonian age, collected from an outcrop on the Ohio Turnpike. Pocket knife is for scale.

It was a pleasant trip across Indiana and into Illinois—until they got to Chicago. The traffic around Chicago was horrible, even by standards of the Washington Beltway. But they managed to skirt most of the city and by the time they reached Rockford, Illinois, things became bearable again. From there it was on to Wisconsin and Minnesota, north to Fargo, North Dakota, and finally into Canada south of Winnipeg.

One of the things that made that part of the trip interesting was the glacial topography that they were driving through. One of the classes that the young man and woman had taken together in college was geomorphology—the study of landscapes—and the professor had spent his career studying glacial terrains. So, while they had heard and read a lot about glacial topography, neither of them had actually seen the distinctive landforms characteristic of glaciated terrains. Or rather, they may have seen them when they were kids, but neither had recognized what they were looking at. In Wisconsin and Minnesota in particular, the landforms characteristic of active glaciation were everywhere to be seen.

They stopped to camp for a couple of days at Maplewood State Park in western Minnesota, partly because of the spectacular glacial landscape. The park was interesting, consisting of lakes (eight of them) in depressions scoured by moving ice, and rounded hills formed by the marginal moraines (sediments pushed aside and dumped by moving glaciers). But it was also beautiful, and it was a lovely place to camp. The young man and woman had meant to just spend a single night there before driving on. But it was such a nice place that they ended up spending three days, hiking the moraine hills and fishing in the glacial ponds (Figure 22.2). It was early summer, the weather was clear and warm, the grass and trees were green, and the ponds were a gentle blue. It was hard to imagine that just 20,000 years ago, this idyllic spot was covered by hard, cold ice that was more than two miles thick.

Figure 22.2—Glacial moraines and lakes at Maplewood State Park,

Minnesota.  Photo by Gail Rosenblum/Star Tribune.

We, or at least the folks that live in glacially sculpted landscapes like Minnesota, are used to the idea that glaciers periodically covered much of North America and Europe in the relatively recent past.  But, of course, that wasn't always the case.  The way that people eventually figured out the reality of the ice ages is one of the more unlikely stories in the history of geology.  That's partly because of the role a poet played in the discovery process.

--------------------------------

Johann Wolfgang von Goethe (1749-1832) is widely considered to be Germany's national poet, and he is certainly one of the most important writers in the German language.  His most famous work, a tragic drama in two parts entitled "Faust", has been termed "*the* drama of the German people".  The script, first published in 1808, revolves around a bargain in which the devil agrees to give Faust everything he wants in this life in exchange for serving the devil in hell for the rest of eternity.  Predictably, this "Faustian" bargain results in Faust seducing a beautiful maiden, a subsequent unwanted pregnancy, and various kinds of murders for which the maiden (not Faust) is ultimately condemned to death.  The tragedy, of course, is that the innocent maiden pays the price for Faust's sin.  The poetic language in German is, by all accounts, hauntingly beautiful.  Unfortunately that beauty does not translate easily into English, and thus Goethe's poetry is less well known in the English-speaking world.  But in addition to Goethe being a famous and revered poet, it also turns out that he was an avid amateur geologist.  As such, Goethe was involved in the discovery that glaciers once covered much of Germany during what we now call the ice ages.[1]

At the age of 26, and on the strength of his first great literary success (a novel entitled *The Sorrows of Young Werther* (1774)), Goethe was offered and accepted employment by the

Duke of Saxe-Weimar-Eisenach. He was promptly appointed to the Duke's Mines and Highways Commission, and Goethe oversaw the reopening of a silver mine. This experience led to his lifelong interest in geology in general and mineralogy in particular. At the time of his death in 1832, Goethe possessed one of the largest private mineral collections in Europe. In recognition of his contributions to mineralogy, the iron mineral with the (approximate) chemical formula FeO(OH) was named *goethite* in his honor.

His role in the discovery of the ice ages is less straightforward. For centuries, people had noticed the presence of unusual and mysterious very large rocks (some as big as a house) strewn across the plains of northern Germany (Fig. 22.3).

Figure 22.3—A typical Erratic Boulder. U.S. Geological Survey file photo.

These rocks, known as "erratics" were clearly different than the underlying sediments or bedrock. It was also noticed that erratics in northern Germany were identical in composition to rocks found in Scandinavia, hundreds of kilometers away. What could possibly have moved them? As it happens, Goethe was also familiar with granite erratics found in Austria and Swizterland which some local mountaineers claimed had been moved by glaciers. Putting two and two together, Goethe surmised that glaciers had in fact transported the erratics that were plainly visible in the Alps. Goethe wrote[1]:

> The glaciers travel through the valleys to the edge
> of the lake carrying the granite blocks loosed from
> above, as still happens today. The blocks remain on
> the lake plain after the ice melts, to be found today,
> unrounded, because they were brought there
> smoothly, and not forcefully.

Based on that quote, which was penned about 1792, one could reasonably conclude that Goethe was the first person to attribute the presence of erratics to the movement of glaciers[1]. Except that that very same idea had been previously published in 1787 by one B.F. Kuhn[2]. Did Goethe get the idea from Kuhn, or did Goethe simply come to the same conclusion independently? In many ways it really doesn't matter because, at least in the short run, Kuhn's and Goethe's idea went exactly nowhere. Most people were simply unaware of what they were saying, and if they were aware they either dismissed it outright or simply ignored it. The day of the glaciers had not yet come.

And yet the basic evidence—the ubiquitous presence of erratic boulders all over northern Europe—was almost impossible to ignore. One of the people who didn't ignore them was a Swiss Engineer named Ignaz Venetz who, in 1816, was working high in the in the Alps attempting to drain an ice-dammed lake. He didn't succeed in that endeavor (the dam failed catastrophically in 1818) but he did happen to notice that the valley showed clear evidence that it had been carved by a glacier that had subsequently retreated. In 1821 Venetz read a paper suggesting that the climate in Switzerland had once been much colder than is was now, and that much of the land had been covered with ice.

At about the same time, in 1815, a Swiss geologist named Jean de Charpentier, an acquaintance of Venetz, happened to be traveling in the Alps. As there no hotels, de Charpentier spent a night in the cottage of a local mountaineer. In conversation, the mountaineer casually mentioned that the glaciers present at higher elevations used to extend much further down the valleys than they did today. The mountaineer said "I find huge boulders of alpine granite perched on the sides of the valleys, where they could only have been left by ice"[3]. That got de Charpentier's attention and he spent several years investigating the geologic evidence that glaciers could in fact move huge boulders out of the Alps and into the surrounding valleys. In 1834, de Charpentier read a paper to the Association of Swiss Naturalists summarizing the evidence he and Venetz had collected, and proposed that in the past, much of Switzerland must have been covered by glaciers. Predictably, these ideas were met with skepticism, but they did intrigue one the scientists in attendance named Louis Agassiz[3].

Agassiz was understandably skeptical of what de Charpentier had to say, but was willing to go look for himself. With de Charpentier as a guide, Agassiz spent the summer of 1836 looking at the glacial terrain near Bex, Switzerland, where de Charpentier directed the local salt works. Agassiz was soon convinced that Venetz and de Chapentier were right, and immediately understood the implications for the geology of not just Switzerland, but all of Europe as well. Importantly, Agassiz was very much part of the European scientific establishment of the day, having studied paleontology with Georges Cuvier and medicine with Alexander von Humbolt in Paris. So, when Agassiz read a paper to the Helvetic society in 1837 essentially repeating what Venetz and de Chaperntier had said earlier, it lent considerable weight to the argument. In addition to his scientific stature, Agassiz's original contribution was to recognize that glaciers could not only explain the landscape of Switzerland, but also the erratics found in northern Germany as well. Furthermore,

Agassiz proposed that, in the relatively recent geologic past, most of Europe had been covered by a vast ice sheet which Agassiz called the "ice period"[4].

Agassiz published these findings in 1837, which was after Venetz had published his book in 1833 and before de Charpentier published his findings in 1840. In any case, once these publications were available, geologists all over the world began to recognize evidence for the presence of "ice ages". Louis Agassiz immigrated to the United States in 1847, becoming a professor of zoology and geology at Harvard University where he continued his studies of glaciers and glacial terrains. When the American geologist Warren Upham identified the presence of a huge ancient glacial lake (much larger than the present-day Lake Superior) that had once covered much of Manitoba, Ontario, North Dakota, and Minnesota, he named it Lake Agassiz. The southern shore of Lake Agassiz, incidentally, is located near what is now Maplewood State Park in Minnesota, where the young man and woman spent an idyllic three days in 1980.

So who gets the credit for the discovery of the ice ages? Would it B.F. Kuhn[2] who suggested that erratic boulders were transported by ice in 1787, or Goethe[1] who expressed the same idea in 1792? We could just as easily give the credit to Venetz, who got his inspiration from his work in the Alps, or de Chapentier who first heard of the idea from a mountaineer in 1815. Or should we credit Agassiz who published a book on the subject in 1837, and whose scientific statue gave the idea credibility?

The answer, of course, is all of the above. In physics, the lonely mathematical insights of a few talented individuals—think Newton and Einstein—certainly led to revolutionary advances. But the earth sciences are fundamentally different. Geology as an avocation or a profession is necessarily observational in nature, and it usually takes a lot of time for new paradigms to be proposed, processed, digested, and finally either accepted or rejected. Continental drift, as plate tectonics used to be known, took a good

50 years from when Alfred Wegener first proposed it in 1912 to when Harry Hess, Frederic Vine and Drummond Mathews demonstrated it in 1963 (Chapter 20). The story of how the ice ages were discovered is exactly the same way.

They all deserve the credit.

-------------------------------

It took the young man and woman another five days to make it all the way to Alberta and her sister's house in Edmonton. Along the way they passed Canadian towns that had, at least to their Americanized ears, exotic-sounding names such as Winnipeg, Regina, Moose Jaw, and Saskatoon. They spent almost no time in Edmonton, preferring to go immediately to Jasper National Park to get their first look at the Canadian Rockies. Unlike the American Rockies, which are composed mostly of igneous and metamorphic rocks, the Canadian Rockies consist largely of layered sedimentary rocks such as limestones and shales. Having been so recently carved by glaciers, these mountains are visually spectacular. The road from Edmonton to Jasper runs alongside the Athabasca River, which now follows a valley once carved by glaciers. The layered limestones, shrouded most of the time by clouds, are beautiful (Fig. 22.4).

Figure 22.4—The young woman and man along the Athabasca River with the Canadian Rockies in the background (1980).

One of the things that made this trip interesting was that in Alberta the glaciers that once covered the mountains have retreated relatively recently in the last couple of thousand years. In contrast, the glacial landscapes they had seen in Maplewood State Park has been ice-free for about ten thousand years. So while the occurrence of glacial hills, valleys, and lakes are similar between Jasper and Maplewood, their appearance is much different. In Maplewood, the glacial moraines are covered with trees and flower-laden meadows (Fig. 22.1). In Jasper, however, the moraines have only recently been uncovered and are largely barren of vegetation. Figure 22.5 shows the young woman at the toe of the Athabasca Glacier in Jasper National Park, which is in the process of retreating up its carved valley, and you can see the barren lateral moraine over her left shoulder.

Figure 22.5—The young woman at the toe of the Athabasca Glacier in Jasper National Park, Alberta. Note the lateral moraine over her left shoulder.

All of this must have seemed obvious to that mountaineer, who in 1815, casually mentioned to de Charpentier that glaciers once filled most of the valleys in Switzerland. After all, it was something that he saw every day. Nevertheless, understanding the significance of those boulders strewn across Germany's northern plain was not at all obvious to most of the people who saw them. But people like the poet Goethe did eventually figure it out.

It just took some time.

## REFERENCES

1. Cameron, D., 1964. Early discoverers XXII, Goethe-Discoverer of the ice age. Journal of glaciology 5(41): 751-754.
2. Kuhn, B.F., 1787. Versuch uber den mechanismus der Gletscher. A Hopfrers fur die Naturkunde Helvetiens (zurich). 1, pp. 119-136
3. Marcou, J., 1886. Glaciers and glacialists. Science vol. VII, no.

181: 76-80.

4. Pering, T. 2009. The History and Philosophy of Glaciology.

# CHAPTER 23.
## ENGINEERING THE EARTH

It took the young woman about an hour on the first day of her new job with Century Engineering to figure out that engineers didn't think like geologists. It wasn't that they were smarter or dumber, nicer or meaner, or more or less knowledgeable. They were just different, and at first she couldn't quite put her finger on just what that difference was.

That morning, her new supervisor, an engineer named Sam Gupta, had begun briefing her on the first project she would be working on. Century Engineering had been contracted to design a building in downtown Baltimore. The first order of business in the design of any structure is to characterize the nature of the soils, sediments, or rocks that lie beneath the site. That information was then used to design a foundation capable of withstanding the weight of whatever was going to be built. This particular building was going to be an office complex about twenty stories tall, so naturally it was important to make sure the foundation could handle the stresses generated by the buildings' weight. That meant drilling into the ground where the foundation was going to be laid, logging whatever soils or rocks were encountered, and reporting this information in a standardized way that the design engineers could understand and use.

So that morning, Sam Gupta was explaining the "standardized" process of recording a drilling log. If soils or sediments were encountered, she was to record their texture (sand, silt, clay) and color. If rocks were encountered she was to log their type (granite, sandstone, shale), count the number of fractures, and again note the color. "Um", the young woman began to ask "don't you want to know what the rocks are made of? What kinds of minerals are there?"

"Nope", Sam answered, "doesn't matter. What matters is how hard they are and how much of a load they can bear". In rock drilling, a diamond-tipped double-tube core barrel was used to spin into the rock, cutting it and trapping the rock core so that it could be brought back to land surface for description. Sam told the young woman not to worry about identifying the minerals in the rock, as she had been trained to do, but rather to focus on whether the rock was fractured or exhibited weathering features. If groundwater was moving through fractures in the rock, it often showed up as a weathering rind around the plane of a fracture. Furthermore, because groundwater can destabilize a foundation, that was something that the engineers needed to know. The boring would continue until clean unaltered rock was reached, which in turn told the engineers how deep the footings of the foundation needed to be set.

Drillers can be a rough bunch, which concerned the young woman as she was driving to the jobsite in downtown Baltimore. But in this case she needn't have worried. Richard, the driller, was about 50 years old and his helper was his twenty-year old nephew Johnny. They immediately began treating her like a kid sister and were very protective of her if other construction workers said or did anything inappropriate. In the 1980s, women on construction jobsites were not a common occurrence. But at this particular jobsite, and under the watchful eyes of Richard and Johnny, harassment wasn't an issue. That issue, however, *would* come up later at other jobs.

What did prove to be an issue was that Richard and Johnny, who liked to spend their evenings drinking in various Baltimore watering holes, were often late getting to the jobsite in the mornings. The young woman, rather than fussing at them, used the morning down time to organize her notes from the previous day and to begin drawing cross sections of the soils, sediments, and rocks underlying the site. That, in turn, was one less thing the engineers back at the office had to do, and so they appreciated her efforts. All and all it was interesting work and, for the most part, The young woman enjoyed it.

The young woman was learning what has come to be called geotechnical engineering. Going back at least to the pyramids of ancient Egypt, the properties of soils and rocks have always been a consideration when building large structures. However, it wasn't until the early 20th century that quantitative methods for measuring the physical properties of soils and rocks were worked out and standardized. In particular, a young Austrian engineer/geologist named Karl von Terzaghi visited the United States in 1912 to study the huge dams being constructed in the west. That experience convinced him of the importance of quantifying the inherent strength and stability of rocks and soils when designing foundations and retaining walls. After spending years developing methods for making such measurements, he published a book entitled *Soil Mechanics* in 1924 which established geotechnical engineering as a distinct discipline within the earth sciences. The young woman's new job was to collect the geotechnical information needed to design and build the foundation of the office building.

Because, when a foundation fails, bad things happen.

--------------------------------

The St. Francis Dam failed catastrophically on March 12, 1928 at exactly 11:57 PM. It was the worst possible time for it to happen. At that time of night, most of the people living along the Santa Clara River were in their homes sound asleep and had no chance to escape. A wall of water fully 120 high and traveling 18 miles per hour crashed down the San Francisquito Canyon destroying everything in its path. When the night was over and the flood waters had receded, 431 people were dead. It was the worst American civil engineering disaster in history.

And it happened because of geology.

Los Angeles is probably the thirstiest city in the United States, if not the world. But it wasn't always like that. When the Mexicans founded a pueblo there in 1781, the Los Angeles River flowed year-round. The first settlers were mostly farmers and they immediately began digging ditches (zanjas) to divert water from the river to irrigate their fields. Interestingly, the first agricultural product for which Los Angeles was known were its high-quality wine grapes. Even after the United States took possession of California in 1848, Los Angles remained an obscure, sleepy agricultural community for the next 30 years, and the Los Angeles River remained the principal water supply. That changed abruptly in 1885 with the completion of the Santa Fe railroad line from Chicago to Los Angles. Initially, immigrants from the east arrived in a trickle, attracted by the agricultural potential and the pleasant climate of the area. But when oil was discovered in 1892 the trickle became a flood, and almost overnight Los Angles was transformed into a boom town. Suddenly water from the Los Angeles River was no longer sufficient for the burgeoning population.

The solution to Los Angeles' water shortage was to tap (some would say steal) water from the Owens Valley 233 miles to the east. A brilliant, and largely self-taught engineer named William Mulholland designed and built the California Aqueduct from the Owens Valley which came on line in 1913. The aqueduct, which was completed on time despite numerous attempts at sabotage by unhappy Owens Valley residents, solved Los Angeles' immediate water supply problem. But at the rate that the city was growing, it was clear new sources of water were going to be needed.

So Mulholland began looking in to building one or more dams in the San Francisquito Canyon to provide water storage for the periods of drought that regularly plagued southern California. He personally made a study of the geology of the canyon, observing that the western wall was composed mostly of a coarse reddish conglomerate interbedded with sandstones that had stringers of the mineral gypsum (calcium sulfate) running through it. Below the conglomerate was an abrupt contact with a mica schist. The contact between the two rock types was abrupt because it was the trace of the San Francisquito fault[1]. The eastern side of the canyon was underlain entirely by the schist (Fig. 23.1).

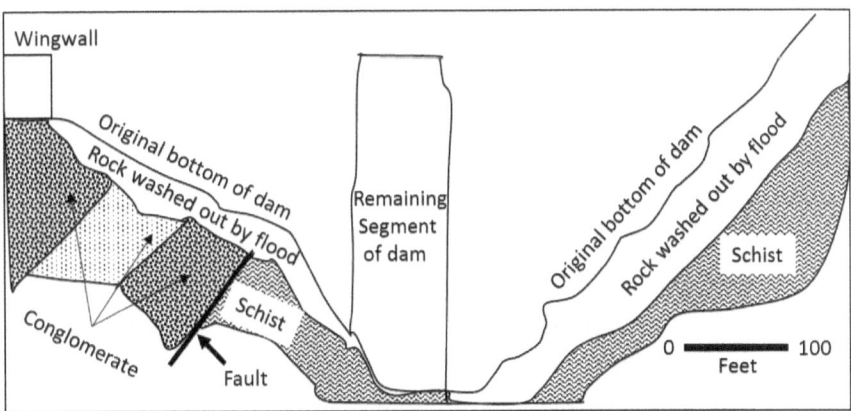

Figure 23.1—The geology underlying the site of the St. Francis Dam. Modified from Ransome (1928)[1].

On the face of it, the underlying geology of the site was promising. The conglomerate was a very hard, tightly-cemented rock and appeared fully capable of bearing the weight of the dam. The schist was somewhat softer, but with concrete footings set deeply into the rock, it too would bear the dam's weight. Construction of the St. Francis Dam began in 1924 and water began to fill the dam on March 12, 1926. In April, the water reached the level of the San Francisquito Fault and, as had been anticipated, water began to seep into the fault trace. Work began immediately to seal the leak off, but this effort was only marginally successful. Finally, a two-inch water pipe was laid to drain the water away from the fault and to help relieve the hydrostatic pressure. All indications were that this had solved the problem.

But there was another problem that no one had anticipated. While the conglomerate and schist rocks had been tested to determine their compressive strength, nobody thought to test how the rocks would react to being inundated by water. While the conglomerate was indeed tightly cemented, it was cemented by the minerals gypsum and calcite.  In the arid west, soil moisture is often wicked to land surface by evaporation, leaving behind whatever minerals happened to be dissolved in the water.  Over time, that results in the soils or sediments becoming tightly cemented by calcite (calcium carbonate) and/or gypsum.  So while the conglomerate underlying the St. Francis Dam was indeed a hard, cemented rock, the calcite and gypsum cements were highly water-soluble.  When the rising waters behind the dam reacted with the conglomerate, the cement binding the rock together began to dissolve, turning it from a hard, competent stone into a mushy gravel.  The dam held for about two years, and then failed.  In retrospect, it became obvious that not taking the mineralogy of the conglomerate's cement, and realizing that the cement would certainly dissolve in water, was the ultimate cause of the disaster[1].

The experience of the St. Francis Dam disaster is one reason why modern construction methods are careful to include an accurate geotechnical description of the materials upon which foundations are built.  That is why the young woman was tasked with drilling into the rock underlying what would become a new office complex in downtown Baltimore, and carefully logging the rock's characteristics according to standardized procedures.

------------------------------

Fritz Meyer was having a hard time deciding on his academic major.  He had come to the University of Maryland in the 1970s with the intention of studying civil engineering, and had dutifully taken the basic introductory engineering courses such as statics and dynamics.  But he also happened to take introductory geology as an elective, figuring that some background in geology would be useful in civil engineering.  The problem was, he got hooked on the geology classes and pretty soon he was finding them more interesting than engineering.  Also, by virtue of taking all those geology classes, he became a good friend of both the young woman and man.

For a while, Fritz considered changing majors, but he was

reluctant to give up on engineering. For one thing, he came from a stolid German family where engineering was held in high esteem. The last thing he wanted to do was disappoint his parents. So for the next couple of years, he simply continued following the sequence of courses leading to both an engineering and geology degree. What made that possible was that the basic science requirements—calculus, physics, and chemistry—were virtually identical for both majors. And so, without really intending to, by his senior year he was in a position to finish a degree in both engineering and geology. In the end, the only course in the geology curriculum that he didn't complete was the year-long senior thesis project. So when he graduated with a bachelor's degree in civil engineering, he was just two credits short of a degree in geology. But as it turned out, having all of those geology classes on his engineering transcript helped him land his first job out of college.

For years going back to the early 1950s, city planners for Washington D.C. had been talking about building a subway system to serve the nation's capital. The Washington Metro Area Transit Authority (WMATA), better known as the Metro, first began construction of the system in 1969. It took a full seven years to complete the first 4.2 mile segment of the Red Line between Rhode Island Avenue to Farragut North, partly because of difficulties imposed by the local geology. The geology of Washington D.C. is incredibly varied, ranging from soft coastal plain sands and clays to hard, fractured metamorphic and igneous rocks. Like it or not, geology was going to be a big factor in building the Metro subway system, and Fritz's background in geology fit the skill set that Woodward-Clyde Consultants, an engineering company heavily involved in Metro's construction, was looking for. Less than a week out of college, Fritz Meyer not only had a job, he had a really good job.

The construction of the Metro subway in the 1970s and 1980s happened to coincide with a profound change in engineering practice when it came to building tunnels. Prior to the 1970s, the accepted practice was to advance a tunnel either by blasting or excavation a short distance, usually less than a meter, and immediately supporting the tunnel with braces and/or bolts to keep it from collapsing. That, of course, seemed to be the safest way to advance a tunnel. But in the 1920s and 30s, engineers in Austria— who were heavily influenced by Karl von Terzaghi's 1924 book *Soil Mechanics*—began to incorporate the inherent stability of soils or rocks into tunnel design. The result was a methodology known as *The New Austrian Tunneling Method*, or NATM that was extensively researched and documented in Austria during the 1950s and 1960s.

That was just before the beginning of the construction of the Metro subway system. And because of the difficulties encountered using the traditional tunneling methods for the initial stretch of the system, it soon became apparent that the cost of building the full system was going to be astronomical. That, in turn, gave the Austrian engineers an opportunity to propose that NATM should be applied in order to construct the Metro system. But because NATM had never been used in the United States before, there were a lot of questions about its efficacy and safety.

NATM is actually more of a philosophy than a single construction technique. In essence, the goals are to:

- Maximize the inherent strength of the soils or rocks,

  allowing for deformation so that the soil/rock becomes part

  of its own support.

- The application of a thin layer of sprayed concrete (known

  as shotcrete) immediately after excavation to prevent

  greater deformation of the soil or rock.

- Systematic measurement of tunnel movements in order to assess the need for further support.

- Using conventional steel rods, steel mesh, or fiber mesh in addition to shotcrete in order to reinforce the structure as needed.

For Fritz Meyer, the most important part of that philosophy was the "*systematic measurement of tunnel movements...*" because his new employer—Woodward-Clyde Consultants—had the contract to instrument the advancing tunnels to assess their stability and/or instability.

NATM is sometimes referred to as "the observational approach" because rather than relying on a pre-existing design for the whole tunnel, it works by carefully observing the behavior of the soil/rock immediately after excavation and using that information to build a custom support system appropriate for that section of the tunnel alone. That, in turn, reduces the cost of construction by not overbuilding or underbuilding the additional support required. In the end, the Washington Metro subway system was the first tunneling engineering project in the United States to make use of the NATM[2]. Not only did it reduce the overall cost of building the subway system, it also made it safer.

Fritz Meyer spent the next couple of years installing instrumentation such as extensometers, devices used to measure movements in rocks as they settled following excavation, and slope indicators designed to quantify the "tilt" of the rocks as the excavations advanced. It was only after a section of tunnel passed a series of stability criteria, with built-in safety factors, that construction proceeded and excavation of the next section could begin. The first section of tunnel that Fritz worked on using NATM eventually became the Wheaton Metro Station. It was excavated into the Wissahickon Schist (Chapter 6) located in the northwestern part of the system about 60 meters below land surface.

------------------------------

As the rock-coring proceeded in Baltimore, the young woman learned about something the engineers called the "Rock Quality Designation (RQD) index". The idea is to try and quantify the degree to which a rock has been fractured, and thus how suitable it is to stabilize a buildings foundation. Even the hardest granites and gneisses are fractured to some degree. In part the fracturing reflects how much movement the rock has experienced over time, with rocks near active or inactive faults being the most fractured. But also as a rock becomes exposed by erosion, simply the unloading of overburden pressure will fracture rocks. More interestingly, it turns out that the earth also moves slightly in response to the moon's changing gravitational pull. These so called "earth tides" can also fracture rocks.

The drillers the young woman was working with were using an NX-size double-tube core barrel 54.7 millimeters in diameter to core into the rock, and the length of the core barrel was 1.5 meters. They were coring into the Baltimore Gneiss (Chapter 19), which is a pretty hard rock, and it often took an hour or more to advance the core barrel 1.5 meters. When the drillers retrieved a core run, they extruded the core into a trough and the young woman examined it immediately, counting the number of fractures and measuring the distance between the fractures. This was tricky work because the coring process itself often broke the rock core, and it was important to distinguish between the natural fractures and the drilling-induced fractures. The RQD index is defined as the sum of the distance between natural fractures divided by the length of the core run, and is expressed as a percent. Thus if a core run of 1.5 meters (150 centimeters) has fractures spaced at 38, 17,29 and 20 cm, the RQD would be:

$$(38 + 17 + 29 + 20)/150 \times 100 = 69.3\%$$

An RQD index of 69.3%, incidentally, is only "fair" for supporting a foundation. For each core hole in the grid pattern they were drilling, the young woman and the drillers were required to keep drilling down until they reached an RQD of at least 90%, which is the definition of an "excellent" RDQ[3].

The use of the RQD index for describing a rock is a good illustration of the different ways that geologists and engineers think about rocks. When a geologist (the young woman) is looking at a rock, the idea is not just to identify what kind of rock it is (granite, limestone, gneiss) but to try and understand something about its history. In the case of the Baltimore Gneiss, that history began with sediments accumulating at the bottom of an ocean prior to about 1.2 billion years ago, probably during the assembly of the supercontinent Rodina[4]. Those sediments were then subjected to at least three different episodes of heating and deformation that turned them into the hard, crystalline rock that is now the Baltimore Gneiss[5]. As The young woman looked at the core samples of the gneiss, she could plainly see the large crystals of potassium feldspar that are known as "augen" (after the German word for an "eye") because, well, they look like eyes. Those large crystals, some as big as her thumb, must have taken millions of years of heating to grow (Chapter 13). That's what a geologist sees when she looks at the Baltimore Gneiss.

The geotechnical engineer (the young woman), on the other hand, is focused on the physical properties of the rock. At what frequency are pre-existing fractures found, do those fractures exhibit evidence of groundwater seepage, is there any observable evidence of weathering that might weaken the rock? All of this is done with the goal of designing and building a structure that will not suffer the ignominious and tragic fate of the St. Francis Dam.

As The young woman got used to her new job, she gradually came to realize just why it is that geologists and engineers think differently. Geologists want to know what a rock *means* in the history of the Earth. Engineers, on the other hand, want to know what a rock can *do* for them when they build things.

Not all the projects that the young woman worked on were as interesting or trouble-free as the Baltimore gneiss drilling project had been. One project involved doing percolation tests in the coastal plain sediments of Anne Arundel County, just south of Baltimore. Percolation tests are necessary to see if the sediments are permeable enough to allow construction of septic fields. A percolation test is performed by digging one or trenches in the proposed septic field to a specified depth, sinking six or eight inches in diameter three to six feet on the floor of the trench, presoaking the holes by maintaining a high water level in the holes, then running the test by filling the holes to a specific level and timing the drop of the water level as the water percolates into the surrounding soil. Empirical formulas are then used to design the size of the septic field.

Perc tests, as they are usually called, are not difficult to perform. But they do require using earth-moving equipment, which in turn means hiring contractors to perform the digging so the geotechnical engineer (the young woman) can do the testing. This being the early 1980s, women were not always welcome on jobsites, and inevitably the young woman eventually ran into that problem. One backhoe operator in particular, a skinny sallow-faced man with yellowing skin, was particularly troublesome. His language was so coarse and filled with sexual inuendo that it was not only offensive, it was literally beyond belief. The young woman did her best to ignore it, as did most of the other men on the jobsite. But after a while it got so bad that one of the construction workers decided to do something about it. Taking the skinny little man aside by the arm and towering over him, the construction worker hissed something in his ear and shoved him back to his backhoe. The construction worker walked over to the young woman and told her the little man wouldn't bother her any more.

"After all", he growled, "I wouldn't let anyone talk to my

sisters like that."

## REFERENCES

1. Ransome, F.L., 1928. Geology of the Saint Francis dam site. Economic Geology, 23(5), pp. 553-563.

2. Rudolf, J., Gall, V. and Wagner, H., 2009. History and recent developments in soft ground NATM tunneling for the Washington, DC Metro. In Proceedings, ITA-AITES World Tunnel Congress.

3. Deere, D., 1988. The rock quality designation (RQD) index in practice. In *Rock classification systems for engineering purposes.* ASTM International.

4. Li, Z.X., Bogdanova, S.V., Collins, A.S., Davidson, A., De Waele, B., Ernst, R.E., Fitzsimons, I.C.W., Fuck, R.A., Gladkochub, D.P., Jacobs, J. and Karlstrom, K.E., 2008. Assembly, configuration, and break-up history of Rodinia: a synthesis. Precambrian research, 160(1), pp.179-210.

5. Muller, P.D. and Chapin, D.A., 1984. Tectonic evolution of the Baltimore Gneiss anticlines, Maryland. Geological Society of America Special Papers, 194, pp.127-148.

# CHAPTER 24.
## THE SECRETS OF MOTHER EARTH

The young man stared at the computer printout that he had just generated and immediately felt his heart sink. The simulation he had just run wasn't even close to the data he was trying to match.

The project he was working on was a study of the hydrogeology and ground-water quality underlying the Baltimore Industrial Area in Maryland. It's often forgotten these days, but the growth of most industrial cities in the eastern United States—Baltimore, Philadelphia, and New York are all examples—were largely driven by the availability of water. The Patapsco River, which feeds Baltimore's harbor, and the Back River to the north were sources of fresh water for the European settlers who first arrived in 1661. However, the quality of the river water was not ideal for drinking, and springs and shallow wells supplied most of Baltimore's drinking water during the 17th and 18th centuries[1]. It wasn't until 1853 that the first artesian wells were drilled in Baltimore. By 1860, there were about 90 wells producing groundwater of exceptionally good chemical quality, most of which were located near what is now Baltimore's Inner Harbor.

The sudden availability of artesian groundwater in the latter 1800s was a boon to the industries that were rapidly growing in Baltimore. By 1896, there were more than 200 wells in Baltimore serving industries as varied as the Standard Oil Company, the Maryland Fertilizer Company, and the Electric Copper Works. Also, because of the groundwater's good chemical quality, there were ten beer breweries and seven liquor distilleries making eager and profitable use of the well water[2]. Then as now, Baltimoreans were a thirsty bunch.

By 1900, these wells were producing about a million gallons of water per day and the rate of well drilling and groundwater pumping continued to increase rapidly. The coastal plain aquifers the wells tapped were highly permeable and individual wells could produce as much as a thousand gallons of water per minute. By the time groundwater use peaked during World War II, the wells in the Baltimore industrial area were producing about 50 million gallons of water per day[1]. But, as often happens when a cheap, easily accessible resource such as Baltimore's artesian aquifers become over-used, bad things can happen. As early as 1920 it was noticed that saltwater was beginning to contaminate some wells located near the Inner Harbor. As pumping increased, more and more wells became contaminated with saltwater, and by 1942 pumping actually began to decrease because of the deteriorating water quality.

The U.S. Geological Survey (USGS) documented this sad history in a report published in 1952, carefully pointing out how the over-use of artesian wells had so badly damaged a valuable resource[1]. By the end of the 1970s, when the young man was looking for his first job, the USGS and the Maryland Geological Survey had decided it was time to reevaluate the groundwater resources of the Baltimore Industrial Area. The idea was to do a three-year project to update the current use of groundwater in and around Baltimore, map the areas where water-quality had been degraded or not degraded, and hopefully give some guidance as to what could be done to prevent further contamination of groundwater.

And the young man got the job.

-----------------------

It's pretty safe to say that the first users of artesian groundwater in the Baltimore Industrial Area in the late 1800s were interested only in one thing—using the water to make money. Whether the water was used to cool copper or steel as it rolled out of the furnaces, or to manufacture natural gas from coal, or just to make beer and whiskey, the groundwater was interesting entirely for what it could be used for. Questions like where the groundwater came from, or why water could be found in some places but not in others, or why the water quality was sometimes good and other times not so good, simply weren't considered.

But Baltimore wasn't the only east-coast city eagerly tapping into the watery bonanza of the Atlantic Coastal Plain aquifers. From Long Island all the way to Alabama and Florida, wells were being drilled using the new truck-mounted cable-tool drilling rigs that became available at the turn of the century. As more and more wells were drilled, and as the sediments being encountered at different depths were logged and described, the geometry of the artesian aquifer system underlying the east coast gradually came into focus[2]. A geologist with USGS named Nelson Horatio Darton spent six years in the last decade of the 19th century traveling up and down the east coast laboriously collecting the logs drillers made as they drilled new wells. Darton soon realized that the underlying layers of gravel, sand, and clay were neatly arranged like the layers of a cake on top of the underlying crystalline rocks. Furthermore, the layers of sediment dipped gently and thickened to the east (Figure 24.1).

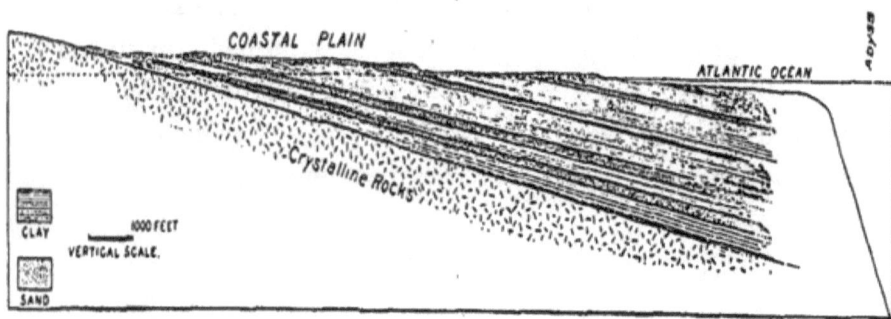

Figure 24.1. Darton's geologic section of the Atlantic Coastal Plain from west to east published in 1896[2].

The hydrology of the artesian aquifers was also relatively simple. Where the layers of permeable sand and gravel are exposed at land surface, they are recharged by rainfall which proceeds to seep downward under the pull of gravity and move toward the Atlantic Ocean. Those sandy "aquifers" are then "confined" by overlying layers of relatively impermeable clays, which effectively pressurizes the aquifers. These pressurized artesian aquifers are prized because, in some cases, groundwater can flow naturally to land surface without having to be pumped. Many of the very first wells drilled into the Atlantic Plain tapped sandy aquifers did in fact flow naturally and continuously. So, not only was the water essentially free (except for the cost of drilling and casing a well), it didn't cost anything to pump it out of the ground either. It's not hard to understand why there was such a rush to tap the watery wealth of the coastal plain artesian aquifers in the late 19th century.

The city of Baltimore is located just on the western feather edge of the Atlantic Coastal Plain, with the northwestern half of the city underlain by hard crystalline rocks (the Baltimore Gneiss) and the southeastern half is underlain by coastal plain sediments. The oldest of the sedimentary layers, the Patuxent Formation, is lower Cretaceous in age (~120 million years old) and consists of course gravels and sands that, near Baltimore, range from 15 to 80 meters thick. These sands and gravels were laid down by a series of swift-flowing rivers that fed a series of coalescing deltas extending out into the Atlantic Ocean[3]. These highly permeable sediments are recharged by rainwater in the high elevations in and around Baltimore and form a productive aquifer named, as you might suspect, the Patuxent aquifer (Fig. 24.2). Many of the first wells drilled in and around Baltimore in the 19th century tapped the Patuxent aquifer.

Then, sometime around 113 million years ago, there was a global sea-level rise that transformed the upland river system recorded by the Patuxent Formation into a vast swamp. This changed the texture of sediments being deposited from sands and gravels to organic-rich silts and clays known as the Arundel Formation, or more descriptively as the Arundel Clay. Interestingly, the terrestrial swamps recorded by the Arundel Clay were home to a variety of reptiles that included dinosaurs and turtles[4,5]. But more importantly for the hydrology of the Baltimore Industrial Area, the silts and clays of the Arundel Formation are relatively impermeable to water and serve to confine and pressurize the groundwater in the underlying Patuxent aquifer, turning it into an artesian aquifer (Fig. 23.2). There are several other aquifers and confining beds that overlie the Arundel Formation, but the historically important source of groundwater in the Baltimore Industrial Area was the artesian Patuxent aquifer.

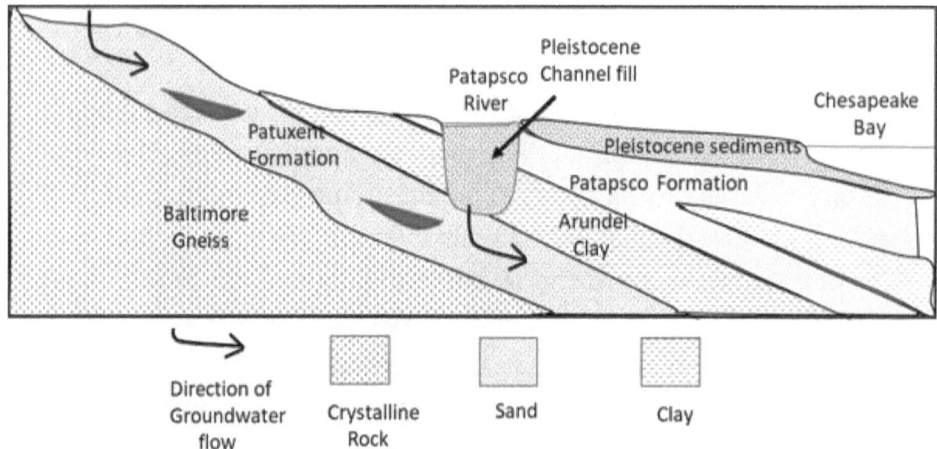

Figure 24.2.—Cross section showing the crystalline basement rocks, the Patuxent Formation, Arundel Clay, and the Patapsco Formation underlying Baltimore.

Part of the young man's job was to construct a numerical model of the aquifer system. In order to do that he, and his co-workers Tracey Kean and Dave Drummond spent months collecting a variety of hydrologic data. They located as many wells tapping the Patuxent aquifer in and around Baltimore as they could find, and carefully measured their water levels. They also took water samples in order to analyze the groundwater chemistry, which later could be used to document the extent of saltwater contamination in the aquifer. Using historical records kept by USGS and the Maryland Geological Survey, they compiled a history of how water levels had changed since 1900 due to all the increasing pumping. They used drillers' logs, electric logs, and gamma logs to measure the thickness of the Patuxent and Arundel Formations across the area. Finally, they used the results of aquifer tests to estimate the hydraulic conductivity, a measure of a sediment's permeability, of both the Patuxent and Arundel Formations.

The young man's job was to then take all of that data, enter it into a computer model that described the movement of groundwater in the subsurface, and then use the model to estimate how the saltwater contamination would move in response to future pumping. The idea was to help city managers and planners decide where to locate future production well fields that would avoid worsening the saltwater contamination of the Patuxent aquifer. After all, it was the unthinking and uncontrolled pumping from 1900 to 1943 that caused the saltwater contamination in the first place.

So, after laboriously loading the hydrologic data into the model, the young man ran it for the first time, hoping that the water levels simulated by the model would at least be close to the water levels he, Tracey, and Dave had measured in the field. But the water levels didn't match. Not only did they not match, they weren't even close. Something was wrong.

And the young man had no clue as to what it was.

------------------------------

For the next couple of weeks, the young man puzzled over the results of his model simulations. He was pretty certain that the thicknesses and hydraulic conductivity of the Patuxent aquifer and the Arundel Clay were accurately represented in the model. After all, he could clearly see the different traces of sands, gravels, and clays on the electric and gamma logs, so getting their thicknesses badly wrong was unlikely. Also, drillers and hydrogeologist had been performing aquifer tests—pumping newly drilled wells and measuring how much and how fast water levels declined in response—for at least the last 70 years. That made it highly unlikely that he had used vastly incorrect values for the hydraulic conductivity of the aquifer and confining bed sediments. It had to be something else. But what?

As he puzzled over this dilemma, he happened to see an announcement of a new USGS Professional Paper by John T. Hack. The announcement caught his eye because John Hack, probably the most respected geomorphologist in the United States at the time, had taught a class the young man had taken a few years before. In addition to being an eminent scientist, Dr. Hack had been a wonderful teacher. He loved his work, he loved talking about it, and (as is not always the case), the students loved hearing about it. One of the things Dr. Hack had been particularly enthusiastic about in the class (the late 1970s) was his work on landscape uplift in the eastern piedmont. Most people at that time assumed that the uplift of the piedmont in the eastern United States had ended with the end of the Appalachian Orogeny 200 million years ago. But Dr. Hack had found evidence that the eastern piedmont had actually been uplifted again in the last 20 million years—a revolutionary idea. The professional paper that was published in 1982 was the evidence for this he had collected over many years.[6]

As the young man read the announcement of Dr. Hack's new paper, it stirred his memory. The young man remembered another of Dr. Hack's papers, published in 1957, which discussed the geomorphology of the Chesapeake Bay estuary. While perusing some the engineering borings made during the construction of various bridges over the Chesapeake Bay Dr. Hack had discovered that during the last ice age (~25,000 years ago), the sea level of the Atlantic Ocean had been about 400 feet lower than it is today. The rivers that fed into the ancestral Chesapeake Bay, therefore, had eroded much more deeply into the underlying sediments than it appears at land surface. By looking at the bridge engineering borings, which logged the kinds of sediments encountered, Dr. Hack showed that the old river channels had eroded as much as 70 meters deeper than they are today. In many cases, these ancient river channels, also known as Pleistocene channels since they formed during the Pleistocene Epoch (2.6 million to 11.7 thousand years ago), had eroded completely though several clay beds and replaced them with much more permeable riverine sediments as sea levels rose after the last ice age. Could that have happened to the Arundel Clay?

The young man sped over to the Johns Hopkins library to look up Dr. Hack's paper. Finding it, he tore through it anxiously. Sure enough, the paper clearly showed the effects of the old erosional channels in bridge borings in fourteen different locations in Virginia and Maryland, and it documented that the channels had been filled in with relatively coarse younger sediments[7]. So far, so good. Three of the bridge borings were located in Annapolis, fifty miles south of Baltimore. But none of the borings were in Baltimore.

The young man knew that if the Arundel Clay had been breached by the ancestral Patapsco River, and the old channel had been filled in with more permeable sediments, it could affect the groundwater circulation patterns in the Patuxent aquifer. Specifically, the old channels could drastically increase the amount of water seeping out of—or into—the Patuxent aquifer underlying the Patapsco River. Curious, the young man returned to his model input files and modified them to increase the hydraulic conductivity in places where the Arundel Clay may have been breached under the Patapsco River. Then he ran the model again. This time, the simulated water levels were much closer to the measured ones. Could that be the answer?

Possibly. But there was one very big problem. In the absence of physical evidence, such as Dr. Hack had provided with his engineering borings, the young man would just be guessing. And since none of Dr. Hack's bridge sites had been in Baltimore, there was no published evidence.

He was still stuck.

---------------------------

The young woman was adjusting nicely to her job with Century Engineering. The people she was working with were reasonably pleasant, with one or two notable exceptions, and the projects they were doing were interesting. One of the more interesting projects had to do with the new Fort McHenry Tunnel that was in the process of being built in the early 1980s. Years ago, in 1957, a tunnel had been built that funneled traffic underneath Baltimore's Harbor making it much more convenient to travel south from Baltimore toward Annapolis and Washington D.C. But over the years the volume of traffic had increased so much that a new tunnel was needed. So, beginning in the 1970s, plans were drawn up to build the Fort McHenry Tunnel next to the older Harbor Tunnel. The construction for the new tunnel was actively proceeding when the young woman began working for Century Engineering, and she was peripherally involved in some of the geotechnical work in support of the construction.

Later that evening over dinner, the young man told her about his dilemma. He could make his model match the water levels in the Patuxent aquifer *if* he assumed the Arundel Clay had been breached by the channel of the Patapsco River during the last ice age. But it all rested on an assumption, and without proof, he would be on shaky ground. The young woman thought for a moment and said "Century Engineering would certainly would have drilled test borings in the Patapsco River for the Fort McHenry tunnel, just like they would have for a bridge. I can ask at work if they have them somewhere." The borings would have been used during the design phase, which would have been several years ago, and it wasn't a given that they would still be around.

But they were there. When The young woman asked Sam Gupta, her boss at Century Engineering about them, he just nodded and gestured toward the map room. "They're in there somewhere", he said, "see if you can find them". It took a couple of hours, but by lunchtime she had found two cross sections of borings across the Patapsco River where the new tunnel was now being built. When The young man looked at them that evening, he could clearly see the Arundel Clay present on either side of the Patapsco River, and he could see that the Arundel Formation was eroded and replaced by channel-fill sediments in the middle of the river (Figure 24.3). Furthermore, it appeared that in one of the cross sections (B-B', Fig. 24.3), the Arundel clay had been completely breached.

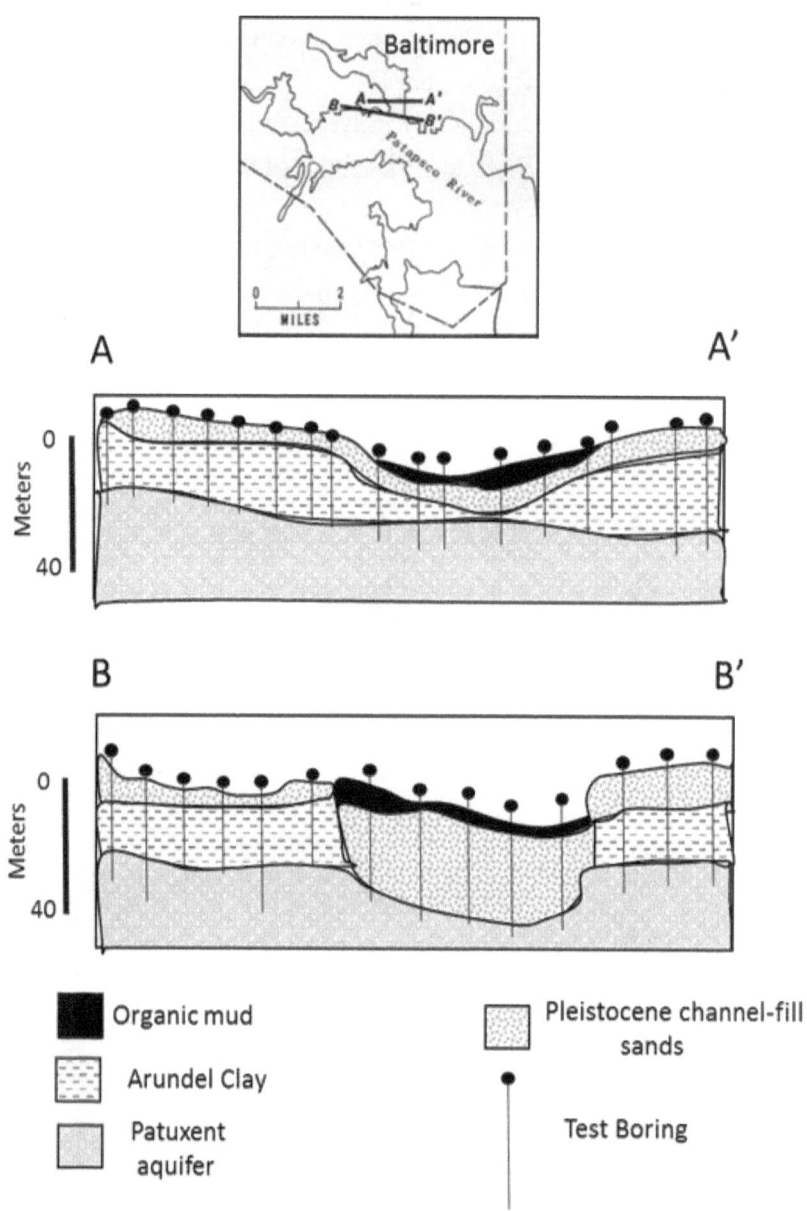

Figure 24.3.—Geologic cross sections at the location of the Fort McHenry Tunnel showing how a Pleistocene erosional channel has breached the Arundel Formation beneath the Patapsco River[8].

That was a promising start, but it got even better. The sediment cores from the engineering borings are typically kept on hand during the construction phase of a project for at least a couple of years before they are discarded. When the young woman asked Sam Gupta about the cores, he told her to look in one of the storage rooms in the geotechnical division building. Amazingly, the people who maintained the core room were just beginning to clear out the old material from the Fort McHenry Tunnel Project. When The young woman asked if she could have some of the core material, they said sure, they're going to be thrown out anyway. So, in addition to having the engineering cross sections, the young man could actually check the sediments to verify that indeed the Arundel Clay was present in some of the cores, but absent in the cores in the middle of the Patapsco River. So, just like John Hack before him, the young man had proof that Pleistocene channels had indeed breached the geologic units beneath the Patapsco River, and that assigning higher hydraulic conductivity values in the model was justified[8]. After that, finishing the model was relatively easy.

Another mystery that the discovery of Pleistocene erosional channels explained was why saltwater had so easily intruded into the Patuxent aquifer in the 1920s and 1930s. Prior to the initiation of pumping in the Inner Harbor District, the erosional channels were a conduit for groundwater to leak *out* of the Patuxent aquifer. That was because the water levels in the aquifer were higher than the level of the Patapsco River. But as pumping increased, water levels in the Patuxent aquifer were lowered below the level of the brackish Patapsco River, and saltwater began leaking the other way, down into the aquifer.

The young man's experience illustrates one the most valuable—and often most overlooked—uses of modeling natural systems like the Patuxent aquifer. Building models, any model, requires making a lot of assumptions that may or may not be justified. If the model containing those assumptions can accurately reproduce historical records, then that suggests those assumptions may be justified. But it's often more informative when the model *fails* to reproduce historical records. That's usually telling you that something is going on that you don't know about. In the young man's case, the "failure" of his initial model pointed out the fact that something important was going on, in this case the presence of Pleistocene erosional channels leaking water out of the Patuxent aquifer, that he wouldn't otherwise have known about.

Sometimes failure can help lead to success.

------------------------------

The last six months of the young man's three-year project studying the hydrogeology of the Baltimore Industrial Area was devoted exclusively to writing the report, getting it reviewed by colleagues, and revising it according to those reviews. In a lot of ways, that was the hardest part of the project. Doing fieldwork was often challenging—locating, drilling, logging, and sampling wells could be hard work—but it had the great virtue that it was done outdoors and away from the office. Writing, editing, and revising reports meant spending long, boring days behind a dreary desk. One thing, however, made those six months bearable.

*The young man and woman had discovered something new.* The Pleistocene channels that had eroded and cut through the Arundel Clay, the way those channels affected groundwater circulation, and how they provided conduits for saltwater to contaminate the groundwater, had existed for thousands of years. But until the young man's model suggested their presence, and until the young woman found the proof of their existence, those channels had remained a secret hidden deep within the earth. In the overall scheme of things, the discovery of those channels was perhaps a small thing. But it did finally explain why the saltwater contamination happened in the first place, and more importantly it suggested ways to prevent the contamination from spreading in the future[8].

Geology, like all the sciences, is often viewed in the context of Great Discoveries like mass extinctions, plate tectonics, or asteroid impacts.  But in truth, progress in geology more commonly reflects steady, incremental advances made by many people who, like the young man and woman, stumbled upon something that at first had simply been perplexing.  But by following the strings of a few leads, they had managed to learn something had never been known before.

The pleasure of that experience is indescribable.

When the young boy in West Point ran out of his house looking for rocks that he could make into arrowheads, but instead found a shining fragment of mica that he mistook for silver, it began a lifetime of noticing rocks.  In the same way, when the young girl crawled to the top of the Rock of Mattapan to play with her siblings, she couldn't help but wonder why it looked the way it did and where it had come from.  She would have been astonished that The Rock had taken a 10,000 mile journey from the southern hemisphere, eventually to crash into North America.

That's what motivates people to delve into the secrets of Mother Earth.

## REFERENCES

1. Bennet, R.R. and Meyer, R.R. 1952. Geology and ground-water resources of the Baltimore area.  Maryland Department of Geology, Mines and Water Resources Bulletin 4, 573 pp.

2. Darton, N.H., 1896. Artesian well prospects in the Atlantic Coastal Plain region.  U.S. Geological Survey Bulletin 138. US Government Printing Office, 232 pp.

3. Hansen, H.J., 1969. Depositional environments of subsurface Potomac Group in southern Maryland. AAPG Bulletin, 53(9), pp.1923-1937.

4. Krantz, P.M., 1998. Mostly dinosaurs: A review of the vertebrates of the Potomac Group.
Lower and Middle Cretaceous Terrestrial Ecosystems: Bulletin 14, 14, p.235.

5. Stanford, R., Weishampel, D.B. and Deleon, V.B., 2011. The

first hatchling dinosaur reported from the eastern United States: Propanoplosaurus marylandicus (Dinosauria: Ankylosauria) from the Early Cretaceous of Maryland, USA. Journal of Paleontology, 85(5), pp.916-924.

6. Hack, J.T., 1982. Physiographic divisions and differential uplift in the Piedmont and Blue Ridge. U.S. Geological Survey Professional Paper No. 1265, 55pp.
7. Hack, J.T, 1957. Submerged river system of Chesapeake Bay. Geological Society of America Bulletin, 68(7), pp.817-830.
8. Chapelle, F.H. and T.M. Kean. 1985. Hydrogeology, digital solute-transport simulation, and geochemistry of the Lower Cretaceous aquifer system near Baltimore, Maryland. Maryland Geological Survey Report of Investigations No. 43, 120 pp.

# AFTERWARD

It's hard not to be overwhelmed by the sheer beauty of the Earth.

Viewed from the Moon, the contrast between the dark blackness space and the soft blues and whites of Earth is so striking that it's no wonder that people, like the astronauts who took this picture, could only gaze in awed reverence. But one need not go into outer space to experience that awe. Standing on the south rim of the Grand Canyon, driving through Rock Creek Park in Washington D.C. on a Sunday morning, or gazing out over South Dakota's Badlands can inspire just as much awe.

The beauty of the Earth runs deeper, much deeper, than the wonderful vistas visible on her surface. Consider the dull browns and grays of the moon in the picture above. That is what the Earth looked like four and a half billion years ago. However, while the moon literally froze in time, remaining virtually unchanged for billions of years, something very different happened on Earth. The accumulation of water into ocean basins, the movement of tectonic plates driven by radioactive heat in the Earth's core, the slow growth of continents from cycles of volcanic eruptions and subsequent erosion, the collisions between those continents, and most importantly of all the emergence of microscopic life 3.5 billion years ago, gradually morphed the dull, gray face of the early Earth into her present blue magnificence. That that metamorphosis happened at all is astonishing enough. That living organisms—human beings—can begin to comprehend *how* that happened is even more astonishing.

But that comprehension did not come easily. Rocks can seem so ordinary, so commonplace, so familiar, that what they *mean* is literally hidden in plain sight. When James Hutton stood on a cliff face in Scotland and noted that horizontal sedimentary rocks were lying on top of much older tilted metamorphic rocks, he realized it could only mean the Earth was impossibly old. That idea was so new, so revolutionary, so astonishing, so awe-inspiring that it clearly staggered him. It wasn't until the very end of his life that he managed to articulate it:

*We find no vestige of a beginning, no prospect of an end.*

That was an idea whose time had come. This book has been an account of how two people, over the course of their lives, came to understand that the real significance of rocks is not what they are.

But what they mean.

# ABOUT THE AUTHOR

During a thirty-five year career with the U.S. Geological Survey, Francis H. Chapelle published more than 130 peer-reviewed research papers, five books, and received the O.E. Meinzer award in hydrogeology given by the Geological Society of America. His research focuses on how the metabolism of microorganisms affects the chemical quality of groundwater in pristine and contaminated aquifer systems.